TELEVISION RESEARCH COMMITTEE

Working Paper No. 3

TELEVISION AND DELINQUENCY

TELEVISION AND DELINQUENCY

by

J. D. HALLORAN, R. L. BROWN
and
D. C. CHANEY

LEICESTER UNIVERSITY PRESS
1970

First published in 1970 by
Leicester University Press

Distributed in North America by
Humanities Press Inc., New York

Set in Linotype Juliana
Printed in Great Britain by
Blackfriars Press Ltd, Leicester

SBN 7185 1088 7

CONTENTS

FOREWORD

THIS is the third working paper published by the Television Research Committee. This Committee was appointed by the Home Secretary in July 1963, to initiate and co-ordinate research into the influence of television and the other media on the attitudes, values, and behaviour of young people.

Unlike the earlier working papers, this paper reports on work that was actually initiated by the Committee. The research on the television viewing habits of certain categories of delinquent, reported in the second part of this book, was started by the three authors when they were working for the Committee as Research Officers or Research Assistants; it was continued and completed at the Centre for Mass Communication Research, which was established by the Television Research Committee at the University of Leicester in October 1966.

The results of the research were available to the Committee when it was preparing its second progress report, which was published in March 1969; in fact, the results are summarized in an appendix to that report.* Although the first part of the present book was prepared for publication after the final meeting of the Committee, and includes some material which was not available for inclusion in the above-mentioned report, it gives a full account of the sort of consideration the Committee had in mind in preparing its report and recommendations.

The first part of the book is intended to set the research in the appropriate wider contexts. After examining the nature and extent of social concern in this area, research into the nature and causes of delinquency is reviewed. We then go on to look at mass communication research, paying particular attention to research which has addressed itself directly to the media–delinquency question. The second part of the book consists of a report on an exploratory study of the television viewing habits of adolescents placed on probation by the courts. Leads for further research which have emerged from this work are also discussed.

Research such as is reported in this book cannot be carried out without the help and co-operation of many people. We are grateful to all those who assisted us in our work, and we are particularly

* Television Research Committee, *Second Progress Report and Recommendations*, Leicester University Press, 1969.

indebted to those young people in all three of our samples who completed the questionnaires, and to the Probation Officers, teachers, youth leaders and students who made this part of our work so much easier than it might otherwise have been. Our special thanks in this connection are due to Mr G. W. Appleyard, O.B.E., Mr G. R. Chesters, Mr F. V. Jarvis, and Mr P. W. Paskell (Principal Probation Officers in the areas covered by the research), Mr J. A. Ferrigan, Miss M. Evans, and the late Mr W. Stitt Dibden.

As with all our work, we had the full support of the Television Research Committee throughout. The advice given by Nigel Walker, Marie Jahoda, Peter Scott, James Hemming, and James Drever, were a great help to us. We are most grateful to the Home Office; to our colleagues, Philip Elliott, Paul Croll and Peggy Gray, for assistance with coding and statistical computations; for the help and co-operation of Judy Lay at the Atlas Laboratory, Didcot (where the data processing was carried out); to Dr Stanley Cohen of the University of Durham and to Philip Elliott for reading the manuscript and for their helpful comments; and to all the clerical staff, especially Freda Swingler, at the Centre for Mass Communication Research for many hours of hard work at all stages in the study.

JAMES D. HALLORAN
ROGER L. BROWN
DAVID C. CHANEY

December 1969

Part One

TELEVISION AND DELINQUENT BEHAVIOUR

MOST sociologists seem to agree that a social problem may be defined as a situation or condition which disturbs society to the extent that *society* tries to do something to ameliorate the situation.[1] If we accept this definition as it stands, then there is no doubt that television and delinquent behaviour can be regarded as a social problem in the United States, although the position in Britain is perhaps not quite so clear. However, if we amend our definition a little and talk in terms of widely-perceived threats to cherished values accompanied by a feeling that something ought to be done, followed by attempts at collective, remedial action—or about heated controversy and organized attempts to persuade the decision-makers to change their policies and practices[2]—then, without making too fine a point, we can say that we, in Britain, also have a problem on our hands.

In examining the nature of a social problem there are several things that have to be borne in mind, and the first is that it takes more than the 'objective situation' to make a social problem. The condition itself, no matter how shocking to an individual, cannot be regarded as a problem unless it offends the values of a considerable number of people. Moreover, these people must not only regard it as a problem in this way, but they must feel that something can be done about it and that some form of social action will be necessary to effect a remedy. Concern, values, argument, influence, pressure and action are all involved in what is clearly a complex situation. Our subject is particularly complex, because we are not dealing simply with delinquency (which may be defined as a problem), or with television as such, or even with the portrayal of crime and violence, but with a presumed relationship between delinquency and television. In effect, the situation is even more complex than this, as we shall see. Only rarely are the interwoven threads disentangled.

We can, of course, accept that a problem exists, as defined above, without taking sides or without granting validity to any of the positions adopted. In this section we shall examine the nature of the problem, and we shall attempt to see how the expressions of concern square up with the 'objective situation'.

Early in 1969, it was learned that American television was to be the subject of a new study by the Surgeon-General of the United States. This study, which would cost a million dollars, came in the wake of the Hearings of the National Commission on the Causes and Prevention of Violence, which itself sprang from a growing anxiety about the increase in juvenile crime in general and violent behaviour in particular. Much of this delinquent behaviour was blamed on television, and it was claimed that the networks' attempt at self-regulation had failed.

The decision to carry out this new investigation in the United States must be seen against the historical and social background of that country in which, compared with this country, violence figures so prominently. Nevertheless, this is not the first time that enquiries in this area—in some form or other—have been carried out at this level in the United States. As part of a wider Senate investigation into juvenile delinquency, a sub-committee under the chairmanship of Senator Thomas J. Dodd, from mid-1961 conducted a series of public hearings, reviewed research, monitored television programmes, and finally submitted a report on Television and Juvenile Delinquency. It is worth noting, in passing, that the sub-committee recommended that a co-ordinated large-scale research attack should be launched to provide more precise information on the impact of television on juvenile behaviour. Nearly five years after the submission of this Report, it became obvious from the proceedings of the National Commission on the Causes and Prevention of Violence in late 1968, that this co-ordinated, large-scale attack had still to be mounted.

In Great Britain in July 1963, the Television Research Committee had its first meeting. This Committee was established by the Home Secretary to initiate and co-ordinate research into the part that television (together with other media and other influences) might play in the development of young people's social attitudes and moral concepts. This was an unusual committee, with no clear precedent in this country. It was the responsibility of the Home Office, received its financial support (£50,000 per annum over five years) from the Independent Television Authority, and was serviced from a university.

Although there was no mention of juvenile delinquency, crime, violence, or any other form of deviant behaviour, in the terms of reference of this Committee, the original conception was certainly not free from these concerns. Moreover, the life of the Committee was punctuated by questions from those who apparently thought that the main task of the Committee was to give a straightforward answer to the question "Does television cause delinquency?"

That these questions should be asked is not surprising when one takes into account the pre-history of the Committee. In November 1961 the Home Secretary held a conference of representatives of religious, educational, social service and other interests to discuss juvenile delinquency, and in particular the extent to which the incidence of delinquency derives from the general state of society. Arising from this conference, the Independent Television Authority offered to finance research into the impact of television on society with particular reference to its effect on young people. When Mr Butler, the then Home Secretary, announced this in the Commons in March 1962, it was clear from the general reception both in the House and outside that delinquency in general and violence in particular were very much in people's minds. A *Guardian* headline referred to the "Problems of showing violence in a violent world" and Norman Shrapnel, the paper's parliamentary correspondent, wrote: " 'Problem' is surely the correct word, though stronger ones have been used. It all came to a head earlier in the week when complaints were made about brutality on the television screen in front of the children, and the Postmaster-General described a scene on the BBC in which Bill Sykes murdered Nancy as 'brutal and quite inexcusable' ".[3] The *Guardian* leading article on the same day was entitled 'Violence on TV'.

During the Commons debate, questions were asked which showed that some members thought that the problem was wider than television and violence, or even television and delinquency. Miss Alice Bacon spoke of "the whole way of life depicted on television", and this approach appeared to be more in line with the stand taken at the conference of experts in the social sciences, called by Mr Butler and held at Sunningdale in Berkshire in 1962. The conference—which was the immediate precursor of the Television Research Committee—recommended, *inter alia*, that the research to be carried out should *not be primarily concerned with the direct study of the effect of television on delinquency*. It was felt that the scope should be wider and should deal with the part that television plays, or could play, in relation to other influences in communicating knowledge and fostering attitudes. The conference also recommended that a committee should be set up to give further consideration to the whole problem, to initiate and co-ordinate research, and to administer the funds that were made available.

These developments may be seen as representing a compromise between a growing public concern on the one hand, and the more sophisticated approach (to both the role of the media and the causes of deviant behaviour) of the social scientists on the other. But they can also be seen as marking the emergence of a new social problem in our society, although there had been rumblings for some time before this. In September 1961 Sir Robert Fraser, the Chairman of the Independent Television Authority, had found it necessary to write an article in *Time and Tide*,[4] in which he answered the critics by attempting to show—by an interesting if unusual use of statistics—that it was "impossible to blame ITV for the post-1955 crime wave because the crime figures had begun to rise before ITV got into its stride".[5] Moreover, as we shall see later, many of the things that were being said about television in the 1960s had been said earlier about films, comics, magazines and the press.

In looking at the differences between the two countries in their approach to this problem, it would be foolish to ignore other relevant differences such as those in the rates of crime and violent behaviour, both on and off the screen. There can be no argument, whatever the criteria or definitions employed, that on all of these scores there is more violence in the United States than in this country. But, as we have seen, social problems do not arise automatically from objective conditions, and there is some slight support for the idea that today in 1969 the alleged television–violence relationship is coming to be *regarded* as a more serious cause for concern in this country than it is in the United States.[6] Obviously past experience, the contemporary situation, general expectations—in fact the whole historical, social–cultural context—are all relevant in this respect. This point should always be borne in mind when Anglo-American comparisons are being made.

It will have been noticed that although we claim to be addressing ourselves to television and delinquency, most of our comments have been confined to television and violence. That we recognize that there is more to delinquency than violent behaviour, and more to violent behaviour than delinquency, should become clear in the course of this work. The social scientist must make his distinctions, and these should be reflected in his thinking and in his research. But, as we shall see later when we are examining the nature of public concern in this area, it is not always possible to make such clear-cut distinctions. The controversy centres on violence, ill-defined though this is.

In many cases the letters complaining about television violence which find their way into the correspondence columns of the press, or into the in-trays at the BBC, probably conceal more than they reveal about the real nature of the concern. In a changing society these particular verbal manifestations could be but one way of expressing a genuine concern about the way some people feel 'the world is going'. Neither is it surprising that television should bear the brunt of the attack, for this is not the first time that technological innovations in communication have led to concern and anxiety.

It can be comfortable to have an easily identifiable scapegoat. If we feel that problems exist and if we are convinced that we know what causes them, then up to a point we can absolve ourselves. We do not have to ask the more troublesome questions which have to do with our own involvement in or our responsibility for the general situation of which the perceived problem could be an integral part. This is but one possible explanation of the nature of social concern; others such as the provision of an outlet for aggressive feelings, a way of giving meaning to a changing and ambiguous situation, or just simply taking the easiest way out of a difficulty may be equally valid explanations for some people in some situations. We are unable to take the matter further here but there is clearly a need for a thorough study of the nature of this and other forms of social concern in our society.

In reflecting on the early stages of the activities of the Clean-up TV Campaign, and the development of the Viewers' and Listeners' Association, a writer in the *Scotsman* commented that:

> "Ever since the war, Britain has been undergoing a slow liberation of opinion and debate. There is a growing tendency not to take received ideas as being absolute, *ex cathedra* pronouncements. The basic pieties are no longer blindly and uncritically accepted.
>
> "But this liberation has shaken deeply the section of the public which likes to believe in a simple, hierarchical, pyramidal system, capped by 'authorities' who mediate on the good, established values this society requires. And so anything which does not unfalteringly owe allegiance to those assumptions is therefore 'immoral'.
>
> "The easiest target for these would-be censors (they censor by causing risks not to be taken by artists, allowing only a cosy, neutered art, as Professor Hoggart puts it) is broadcasting—and especially the BBC. For many years, the BBC was part of this monolithic structure of society, operating with the church and the monarchy to present a simple, authoritative code of morality. Now

that the BBC recognizes that society is no longer like this, and is allowing a greater directness and honesty of debate, the protesters feel betrayed.

"The BBC, as a 'public servant', is much more vulnerable to attack than ITV, which is condoned because it must be responsive to a profit morality as well.

"Some individual incidents on BBC TV have been indefensible—the schoolboy lavatorial giggles of TW3, the tactical blunder by Kenneth Tynan of using that word at that time, when the protesters were in full cry already. But the BBC is no longer one monolithic, authoritarian voice; it is a plurality of voices, reflecting as faithfully as it can the diverse voices and accents of our changing society."[7]

It should be emphasized that the attacks from these quarters on the BBC, and to a much lesser degree on ITV, are by no means confined to the violence–delinquency area. It is worth noting that media violence seems to have figured more prominently as the movement has developed, and that recently there appears to have been a shift of attention from violence in entertainment to violence in news and current affairs. That this shift coincides with the development of student protest and that violence is defined rather selectively are not surprising when the nature of the concern is taken into account. Mrs Whitehouse, who has been the major force in the campaign and the central figure in the activities of the Viewers' and Listeners' Association, has become a focus for many of these fears and anxieties. If she had had her way, it is said that in addition to quite a number of Wednesday Plays and many *Till Death us do Part*, *Man from UNCLE* and *Dr Who* programmes, *Up the Junction*, *Culloden*, *Meeting Points* that dealt with the new theology, the discussion programme featuring Cohn Bendit and the students, and Edmund Leach's Reith Lectures (on radio), would all have fallen to her axe.[8]

Mrs Whitehouse detects people within the BBC who are "hell bent on the destruction of accepted standards", and she appears to fear a political and ideological conspiracy to undermine the British way of life. Although acknowledging the high quality of much that is provided by the BBC and the integrity of many who work there, the Corporation is nevertheless accused of projecting a "propaganda of disbelief, doubt and dirt" into millions of homes. It is claimed that violence is exploited, and promiscuity, infidelity and drinking presented as normal and inevitable.[9] This illustrates the range of attack and the broad-based nature of the concern and anxiety. It may seem a long way from our starting point, which many will no

Mary White-house

doubt see as having more to do with the boy who sees some violent deed portrayed on television and then imitates this shortly afterwards in real life. That this sort of imitative behaviour is included in the concern is not denied. But in treating television and delinquency as a social problem, in the sense that this has been defined, the part played by these other more broadly-based anxieties and resulting pressures needs to be recognized. This is not to say that all those who feel that violence on television leads to violence in real life, or that television causes an increase in delinquency, necessarily share the ideological beliefs of Mrs Whitehouse and her close supporters. The nature and extent of the overlap is not known, but it could be that it is greater (at least in some areas) than many liberals would like to believe.*

Although minority pressure groups such as the Viewers' and Listeners' Association do tend to give a misleading impression of both the extent and authoritative nature of their support, it would be unwise to underestimate the number of people in this country who in some way or other feel that television is an important cause of crime and violence. A recent national survey carried out for the ITA[11] showed that, out of 12 possible causes,† violence in television entertainment was ranked fifth, being mentioned by 35 per cent of the respondents as a '*very* important cause of crime and violence' (see Table on next page). When those who answered 'fairly important' are taken into account, the figure is 64 per cent. Of the other causes, 'Theatres showing movies with violence and sex', 'Coverage of riots and violence on television news', and 'Coverage of riots and violence in newspapers', were ranked sixth (33 per cent), eighth (29 per cent), and eleventh (21 per cent) respectively.

* Quite a number of Mrs Whitehouse's more notable supporters have been associated with other pressure groups whose campaigns have covered homosexuality, capital punishment, gambling, temperance, immodesty, and Sunday Observance. Some of these causes are not likely to attract many supporters, but it is worth noting that a recent national poll showed 70 per cent in favour of hanging, and 77 per cent agreeing that there was too much publicity given to sex. The reaction against or refusal to accept the values of the so-called permissive society may be more widespread than many people think.[10]

† Respondents were shown a card listing 12 possible causes. They were asked to go through the list and say for each one whether they thought it was a *very* important cause of crime and violence, fairly important, or of little importance. To facilitate a direct comparison, the list of 'causes' was the same as the one used in a similar survey in the USA. This meant that the 'causes' were likely to be more applicable to the American situation. Moreover, the premises or theoretical underpinning of the list of possible causes are not known and any such list as this inevitably structures the response. It should also be noticed that crime and violence are coupled in one question.

Comparisons with the answers to the same question from respondents in the USA show that the greatest difference occurs in the 'Violence in TV entertainment' item. This is ranked eleventh in the USA, being regarded as a very important cause of crime and violence by only 27 per cent of the respondents. In both countries violence in entertainment is seen as more of a problem than violence in news. The war in Vietnam comes bottom of both lists.* Another survey[12] carried out in Great Britain in 1967 showed that 57 per cent of the respondents (53 per cent in 1965) thought that there was too much violence on television.

	Percentage saying 'Very Important'		*Rank order of factor*	
	USA	*Gallup* (GB)	USA	*Gallup* (GB)
Possible causes of crime	%	%	%	%
(a) General breakdown in respect for authority, law and order	74	58	1	1
(b) Use of drugs	68	55	2	3
(c) Laws that are too lenient or not letting the police do their job	64	57	3	2
(d) Bad examples set by parents	60	53	4	4
(e) Conflict between whites and blacks	50	23	5	10
(f) Poverty and poor housing	43	28	6	9
(g) Youthful rebellion	42	31	7	7
(h) Theatres showing movies with violence and sex	39	33	8	6
(i) Coverage of riots and violence on TV news	35	29	9	8
(j) Coverage of riots and violence in newspapers	30	21	10	11
(k) Violence in TV entertainment	27	35	11	5
(l) The war in Vietnam	26	12	12	12

It is appreciated that it is possible to interpret these figures in many ways, and that they are not unrelated to the media's presentation of the issues. Still, when they are taken together with other indications of concern, such as newspaper articles and correspondence, letters of complaint to the broadcasting institutions, and parliamentary debates, it is difficult for the social scientist to dismiss the whole affair as unworthy of consideration.† It would also be a mistake for him to assume that the problem had become a problem solely because of the newsworthiness of the activities of the Viewers'

* We are grateful to Dr Ian Haldane (ITA) and Gallup Polls Ltd for giving permission for these figures to be used.

† It should be remembered that the validity or justification of the concern is not being questioned, at least not at this stage.

and Listeners' Association, or because of the efficiency of that Association's public relations. These undoubtedly played a part in the emergence of the concern as a social problem, but whatever the rights and wrongs of the position, it is now a social problem in its own right. As social scientists working in this field, we need not take it at its face value, or as others might define it for us (it has its latent as well as its overt aspects)—but we cannot ignore it. We have to take it, and deal with it as we would any other social problem.

In attempting to deal with the relationship between television and juvenile delinquency, we have one or two other difficulties to contend with, in addition to those which have just been mentioned. As stated earlier, we are not concerned solely with television, nor solely with delinquency which—despite the controversies about the definition of the concept—is usually regarded as a social problem. We have to look for relationships between the two, but first we need to look at the extent, nature and changes in the patterns of delinquency in our society as well as at the ways in which television can influence people, both with regard to delinquent behaviour and more generally. We shall do this by surveying the research in these and related fields but, before we attempt to do this, it might be useful to look a little more closely at the cases put forward by some of those who claim that there is a relationship between television and delinquency. It is one thing to state that there is a relationship—it is another quite different thing to spell out this relationship. This really means that we have to shift our attention to someone like Frederick Wertham—for it is Wertham, the American psychiatrist, who has had so much to say and write not only about what he considers to be the effects of television but also about his opinions on the effects of films and comics. Wertham as an American writes primarily about the American scene and, as already mentioned, we must make allowances for this, although the general line of his argument transcends national as well as media boundaries.* A study of Wertham's writings over the past 20 years enables us to see the continuities in the anti-media case. Occasionally there are references to the unique and special effect of television, although this is rarely spelled out beyond statements of the obvious. The most interesting aspect of this and other parallel or related work is the similarity in the cases that are brought against the various media.

* There are differences between the two countries all across the board, but it seems likely that the differences will be more marked, or at least more significant, in the case of television than they appear to have been in the case of comic books and films.

Wertham's work is known on both sides of the Atlantic and even further afield. In view of the part he has played in the fight against media policies and practices, his views are worth quoting in some detail: "Comic books are known to be an important contributing factor in juvenile delinquency . . . they indoctrinate children with stereotyped images and prejudices . . . particularly insidious is the harmful influence . . . on the sexual development of children in the direction of sadism, masochism, homosexuality, frigidity, and sexual hypochondriasis."[13]

Wertham does not argue that imitation is automatic, or that every child who reads a comic book is injured in some way or other, but neither does he believe that comics have a bad effect only on the maladjusted or otherwise disturbed child. Normal children are influenced in a different way:

"They (comic books) immunize a whole generation against pity and against recognition of cruelty and violence."[14]

"Looking at such pictures over and over again, if it does not cause anxiety, makes children obtuse and oblivious to the sufferings of other people."[15]

"The most dangerous effect they have is a subtle distortion of human value."[16]

Apparently, some children imitate brutality, others have their feelings deadened. Marie Jahoda summarizes Wertham's points as follows:

"(1) Anti-social impulses, which may be caused by a variety of factors, are reinforced and stimulated by reading.

(2) Constant exposure to tales of violence and horror will destroy sensitivity in the 'good and normal' child."[17]

As the years passed, the targets changed and gradually the cinema and television came under attack. In an article—'How Movie and TV Violence Affects Children'—Wertham repeats his earlier arguments that:

"all children are impressionable and therefore susceptible . . . it cannot be sensibly argued that children who see violence on the screen do not acquire a liking for it at some level of consciousness . . . the child whose memory is filled with screen violence may have less psychological resistance to various evil influences . . . such violence blurs his moral sense . . . violent shows lead children to expect, and in some cases to crave, a kind of violence that they will not encounter in real life unless they stir it up themselves . . . the *modus operandi* in an increasingly large number of crimes committed by young offenders closely parallel TV shows they have seen . . . violence is not the best way to settle human differences—but we seem to be using the marvellous technical media of movies and television to teach children that it is the only way . . . one reason we are making so little progress in the prevention of juvenile delinquency is that we tend to blame it all on 'the family' and ignore those outside forces which bypass the family and work directly on the child."[18]

Wertham does not appear to make any major distinction between the large screen and the small screen, but he does feel that "the moving visual image on movie or TV screen—complete with sound" has a much greater impact on most children than the images they conceive in their own mind's eye from reading a story or having one read to them. He maintains that "live action" works directly on the child without the "cumbersome detour of reading", or without the child having to overcome the barrier of "the very effort . . . to read the abstract letter on the printed page" before merging completely with the story. It is interesting to note that we heard little of these 'detours' and 'barriers' when Wertham's aim was directed at the comics.

Mass media in general and television in particular are seen as a school for violence, a school where children do not learn "the evil in horror and the wrong in violence", and where they have lost their natural sympathy for the suffering of others. The trouble, according to Wertham, is not that they get frightened but that they do *not* get frightened. He summarizes his position as follows: "Whether crime and violence programmes arouse a lust for violence, reinforce it when it is present, show a way to carry it out, teach the best methods to get away with it, or merely blunt the child's (and adult's) awareness of its wrongness, television has become a school for violence."[19]

In more recent years Wertham has, not surprisingly, turned his attention to the media portrayal of the war in Vietnam.[20] He argues not only that television is hardening Americans to the war, but that the media presentation of the war cannot be evaluated in isolation because the audience has already been well prepared by the violent portrayals of television in the past.*

The war and the mass media, particularly television, are said by

* There is, of course, another argument with regard to television's portrayal of the war in Vietnam. It is maintained that the vivid presentation of the horrors of war in the drawing room of the average American has led to a revulsion, or at least to a realization, of what is involved in modern warfare. This in turn, so it is claimed, has had a profound influence not only on attitudes about America's involvement in the war, but on wider political issues including presidential election.

Arthur M. Schlesinger[21] in his examination of America's violent way of life to have helped to give new life to the American propensity to violence. He argues that, although the mass media do not create violence, they reinforce aggressive and destructive impulses and may well teach the morality as well as the methods of violence. Comics receive their share of the blame, but Schlesinger stresses the pervasiveness of television, and states that the electronic media foster the subculture of violence with far more vividness and in far greater depth than the older typographic forms.

In surveying just a few of the expressions of social concern about the possible relationship of television to delinquent behaviour in general and to violent behaviour in particular, we have moved from England to America and from the type of concern normally associated with Mrs Whitehouse to that expressed by Wertham and Schlesinger. All three are worried about what they feel television is doing to a valued way of life; all three (especially the two Americans) pay attention to news and current affairs, particularly to the war in Vietnam, as well as to drama and other forms of entertainment; but all three do not subscribe to the same wider value position.

Like some of Mrs Whitehouse's more notable supporters, Frederick Wertham has been a supporter of other causes too. He is known for his social concern in several areas but in some of these other areas it is not likely that he would find himself allied to people like James Dance or Harold Gurden. Neither is it likely that Arthur Schlesinger would have much in common with Sir Cyril Black or Mr James Dempsey.*

This is not a matter that we can pursue further here but, as indicated earlier in discussing the nature of social concern, it is important to bear in mind the different backgrounds, interests and value positions from which the anxieties and criticisms stem. Patterns of media ownership and control, historical background, the contemporary situation and the general social-cultural context, are all relevant. The same sort of things can be said about the influence of television, the same type of attack mounted, for different reasons in different places, or by different people in the same place.

We have found a fairly wide variety of assumptions about the nature of the effects of television. These, some more central than others, include the deadening of sensitivities, a lowering of standards and values, a presentation of misleading pictures of what the world is about, a legitimization of deviant behaviour providing the glamour

* All of these are MPs who have been associated at some time or other, in some way or other, with the Viewers' and Listeners' Association.

of publicity for criminals, an increasing willingness to accept what is wrong, evil and violent, a dissemination of techniques, and a direct causing (e.g. by imitation) of delinquent or some other form of deviant behaviour.

Although these assumptions are those that appear to us to underlie the violence debate, they clearly apply to other forms of delinquent behaviour outside this single field. In a sense, as we have seen, it is not a single—or at least an isolated—field. Granted present circumstances, an understanding of the nature and extent of the social concern in this vital area of delinquent behaviour cannot fail to broaden and deepen our understanding of the problem as a whole.

There is one other area of possible effect about which some people (on the whole, not those referred to earlier) have expressed concern. This concern is about the alleged consequences of the heavy emphasis in the mass media (particularly, but by no means solely, through advertising) on wealth, prosperity and material goods as symbols of success, without a corresponding emphasis being given to the legitimate or approved ways of achieving these goals. It is said that although a great deal of advertising aims to produce discontent and to awaken or even to create needs, the media have nothing to offer with regard to the satisfaction of these needs. It is claimed that the media not only stimulate but constantly remind people, particularly poor people, of their relative deprivation, thereby almost inviting them to go out and help themselves.[22]

Compared with the manifestations of concern about the media portrayal of violence, voices are relatively muted on this last-mentioned score. In view of what has been written already about the complexity of factors underlying the concern, this is not surprising. It may, however, surprise the observer from another planet when—after studying the criminal statistics—he discovers that most offences are property offences, and that most of these are committed by those in the lowest socio-economic groups. However, as we said earlier, it takes more than the 'objective situation' to make a social problem.

This brings up to a further point which, although connected with, and in fact implicit in, much of what has already been written in these pages, requires closer scrutiny before we conclude this discussion on the nature of social concern. We refer to something akin to what Marshall Clinard probably had in mind when, under the heading *The Newspaper and Crime*, he wrote: "The press has been charged with generally promoting and glorifying crime because of the volume of its news items . . . The amount and prominence of space devoted to crime in the newspapers and the amount of con versation based on these stories present a bewildering picture of immorality in our society. By continually playing up crime, it is likely that newspapers are important in making us a crime-centred culture. As a result crime often seems more frequent than it is."[23]

What people read in the papers and see on television certainly influences their views about crime and law enforcement. Indeed, as Stanton Wheeler[24] has reminded us, one of the earlier studies in this area showed how public estimates of the amount and type of crime in the community were more closely related to newspaper reports than to the actual amounts of crime as recorded by the police. We have written at some length about the several motives and anxieties which underlie the social concern, but we must be careful in case in doing this we fail to give adequate attention to core factors, namely, that there is delinquency in our society, that it does appear to be increasing, and (this is the important point at this stage) that most of us rely on the press and television for our information about the nature, causes and extent of this delinquency.

What pictures are being presented to us? Are they false pictures in which the extent of delinquency is exaggerated and its nature distorted? Is it true that by designating crime as news (the more deviant, the more newsworthy), a climate of alarm is created, and fears, anticipations and expectations about delinquent behaviour built up? Does the so-called objective reporting and impartiality lead to a situation where delinquency is seen simply as a matter of interest, but not one of moral concern?

These important questions bring us to the even more fundamental ones of: What is news? Do the mass media create new 'facts' by making non-news news? Must the negative, deviant or sensational predominate? And so on. These are questions which cannot be pursued here, but we shall return to them—albeit briefly—later in this Part.[25]

What we can say at this stage is that, on the whole, the way the news about delinquency is presented by both press and television means that it is unlikely that the facts* will be placed in meaningful context or that the issues behind the story with regard to offence, offender, victim or official agency will be adequately covered. As Wheeler says:

"When criminal events are reported in the mass media, it is of

* This also raises the whole question of the way in which criminal statistics are collected and presented. The media cannot do much about this, except perhaps lend their support to a campaign for the development of "consumer-oriented crime statistics" which would have more real meaning for the public.[26]

course natural to focus on the immediate case. That is where the drama lies, where the action is, and what the public wants to know about. But in the nature of the case these materials provide a poor basis for the development of a full and rounded understanding of crime. The formation of sound social policy typically depends on knowledge of changes in the rate and distribution of the relevant events, but policies more frequently are formed in reaction to certain extreme cases . . . It should be possible to develop creative ways of reporting that would show how and where a particular criminal act fits into the broader picture of the nature and distribution of crimes. To do so would be to encourage a more responsible public opinion."[27]

Admittedly, this raises the problem of the function of the press and television. Granted the commercial system within which the media operate, readers and viewers have to be won and kept. Crime and delinquency as normally presented make a valuable contribution in this connection. Do we have to wait for a change in the system, a new set of news values, before we can hope for a more balanced and realistic picture and a more responsible public opinion with regard to delinquency in our society?

What do we really know about the extent, nature and changes in the patterns of delinquency in our society? To attempt to answer this question in full is beyond the scope of this book, but it is a question that must be asked and to which we must attempt some answers before we go on to discuss what research has established about the effects of television, the nature of delinquent behaviour, and the relationship between the two.

It is not an easy question to answer. In his comprehensive study of crime and punishment in Great Britain, Nigel Walker, the Oxford criminologist, writes:

"Anyone who attempts to assess the relative prevalence of different types of delinquency, their distribution among various subdivisions of the population, and their temporal trends, has to acknowledge peculiar difficulties. Not only is he uncertain how much of any given type of conduct is being reported and recorded, but also, since criminal statistics at present use legal definitions as the basis of subdivisions, he must recognize that even what is recorded under the same heading is a very mixed bag. He is in the position of an ornithologist who is trying, for example, to map the distribution of wild birds by using reports from people who can distinguish birds only as large or small, aquatic or non-aquatic, migratory or non-migratory."[28]

D. J. West,[29] in his book *The Young Offender,** makes a similar point when he argues that changes in the methods of classifying and recording crimes could make the official statistics much more realistic than they now are as measures of crime trends. Police efficiency, changes in police policy, increased public awareness, and changes in methods of reporting and recording, as well as changes in the law, make it very difficult to say anything very definite about changing trends and patterns in delinquent behaviour. As West points out, the fact that the total of convictions rises inexorably year by year could mean no more than that each year a little more of a virtually limitless reservoir of crimes is tapped off and enshrined in official statistics.

The overall figures although not the annual rate of increase have shown a clear upward trend for many years now, but talk of a crime wave has been with us even longer. We seem particularly keen to establish the waywardness and lawlessness of youth, and this is coupled with a sustained and vigorous attempt to show that it was not like this in the good old days. We are not well served by our sense of history: "Even according to official statistics, though fashions in crime may change, people are not necessarily in all respects more lawless today than they were in the last century."[30]

But surely, it will be argued, it cannot be denied that in recent years convictions for crimes of violence have not only been increasing rapidly, but the increase has been more rapid than for other crimes, and more marked amongst young people than amongst other sections of the population. Walker[31] argues that, because of the reasons already mentioned (changes in reporting and recording etc.), the figures on the overall increase in crimes of violence against the person should not be taken at face value. He does not seek to dismiss the whole of the increase as a statistical artefact, or as the product of some other external change. He accepts that some of it almost certainly reflects a real trend. He is quite certain, however, "that the real trend is not nearly as spectacular as the statistics make it seem".

Writing on the nature of vandalism in the sixties, Stanley Cohen[32] states that there is no doubt that the extent of the increase in vandalism, as well as its seriousness, in the peak period 1962-7, was grossly exaggerated. He accepts that vandalism has a very low detection rate, but shows that the percentage increase in the amount of vandalism recorded in the criminal statistics since the beginning of the sixties was not excessive, and certainly not disproportionate

* These passages on statistics and trends rely heavily on this work.

compared with other juvenile and adult offences. The figures do not indicate any dramatic increase, and they show how the very small contribution of vandalism to the total crime picture did in fact decrease between 1961 and 1967. Cohen also argues that: "Nothing in these figures contradicts the point that the mass media inflated the increase in vandalism during these years".[33] He accepts, however, that taken in themselves the statistics do show a pervasive problem.

With reference to the increase in crimes of violence over a longer period, West points out that a significant part of the increase is only apparent, and is really due to the policy of charging as indictable offences acts which used to be dealt with as non-indictable common assaults. Drawing on McClintock's[34] work on crimes of violence, he suggests that there is some confirmation of the common observation that violence committed by young persons often takes the form of rowdyism and fighting around cafés, dance halls, and similar resorts frequented by teenagers. In the wider field of general delinquency, it is also worth noting that many of the crimes recorded have to do with offences of dishonesty and many of these, according to West, consist of comparatively petty and unimportant incidents.

However, when one looks at the criminal statistics[35] and sees the graphs and figures which deal with crimes of violence against the person, one cannot fail to notice the steep and disproportionate rise that has occurred over the past ten years. In spite of all the qualifications that have been made, the fact that in 1968 over 18,000 people were found guilty of indictable offences involving violence against the person does appear to present a problem in itself. Yet these offences must be placed in perspective. Violence, including robbery and assault, and non-indictable assaults, accounted for around 2 per cent of all people found guilty of offences in English courts in 1968. Sexual offences amounted to less than 1 per cent. For indictable offences only, the figures are just over 7 and 2.5 per cent respectively. However, violence against the person now ranks eighth, whereas ten years ago it ranked twelfth (its share of the total having increased from 0.8 per cent to 1.2 per cent) in the table which shows people found guilty of offences of all kinds. As Nigel Walker points out, the types of crime about which most concern is expressed are among the least frequent. Traffic offences account for nearly two out of three of all cases, with property offences and drunkenness a long way behind in second and third places. People found guilty of larceny, breaking and entering, receiving etc., make up over 80 per cent of all those found guilty of *indictable* offences. According to Walker, if we had the true picture of all criminal activities, property offenders would be even more in evidence than they are now. Sexual offenders would also figure rather more prominently in the tables, for both these offences have a low detection and reporting rate. It is worth noting that — in all probability — "in this corrected perspective, violence would appear even more insignificant".[36]

In an attempt to set the delinquency problem in perspective, West writes in *The Young Offender*: "A cool scrutiny of the statistical and historical evidence suggests that the youthful crime situation today is neither unique nor so serious as commonly supposed. Systematic enquiries have demonstrated time and again that nearly all youths commit occasional delinquencies, especially acts of dishonesty and social defiance. There is no reason to suppose this is a new development".

On the whole, when we are generalizing about delinquency, we are talking about thieves, for delinquency consists mainly of property offences committed for the most part by youngsters who fairly quickly grow out of their delinquent habits. As far as attitudes and general behaviour are concerned, most of them are not particularly different from their fellows, many of whom have had the good fortune not to be caught. The fact that most of them are so ordinary probably means that their delinquent response is a fairly normal one for their age, granted their social background* and the general social conditions. The increase in convictions cannot be divorced from the increase in opportunities and temptations, which are an important aspect of these social conditions.

In conclusion, it would appear that most of the experts in this field, although reluctant to comment on the real extent of the delinquency problem, accept—notwithstanding all the qualifications—that there has been a genuine increase at least in certain kinds of delinquent activity (including some forms of violent behaviour), and that this increase is particularly marked in the 15- to 20-years age range. These trends and patterns, which the official statistics probably and the media almost certainly distort and exaggerate, are common to most urban-industrial societies in Europe and North America.†

* This analysis is restricted to 'working-class delinquency' which, although middle-class delinquency is increasing, accounts for a very high percentage of what is normally referred to as juvenile delinquency.

† Denmark is one of the countries where these trends are not so apparent. It has been suggested that this is not unconnected with the greater permissiveness in that country with regard to sexual behaviour. It is doubtful if a 'solution' along these lines would have the support of many of those who are concerned at the present state of affairs in this country.

Having studied the nature of the concern and having examined the trends and extent of delinquent behaviour, we must now begin to consider the relationship between delinquency and television. We propose to do this in two stages, and the first of these involves looking at what social scientists have found out about the nature and causes of delinquency. We shall then turn to the work that has been carried out on the effects of television, special attention being given to research which has addressed itself directly to the media–delinquency relationship.

We are by no means the first to attempt this task. As we have seen, the concerns which are now being expressed about the possible influence of television have previously been expressed about the influence of films, comics and other media. Moreover, up to a point, each wave of concern has been accompanied by a parallel enquiry. At first sight, it could appear to the man in the street that the so-called experts are not so expert at providing answers to what—at least on the surface—appear to be relatively simple questions. However, perhaps we are already beginning to see that the situation is not quite so simple as is normally supposed. It should also become clearer in the course of this work that we are not likely to improve our understanding in this area unless we ask the right sort of questions. We shall discuss later whether the research has in fact always posed the right questions.

There is a vast literature on delinquency and its causes, and we cannot hope to cover it adequately in this book. However, we shall make a brief survey of what in our opinion are the main approaches to the problem, so that we can obtain a good idea of what criminologists, psychologists and sociologists have established about the cause of or factors associated with delinquent behaviour. Naturally, we shall be very much on the lookout for any mention of media influences.

As it happens, this part of our search is not likely to be very rewarding. In the vast literature on delinquency, the role of the media does not figure prominently, in fact at times it does not even earn a mention. Taking a few well-known general books, we find that Walter Reckless, in *The Crime Problem* (USA, 1955),[37] does not refer to the media when discussing basic issues in the causation of crime; neither does Howard Jones in *Crime in a Changing Society* (England, 1965);[38] nor Nigel Walker in over 350 pages of *Crime and Punishment in Britain*, also published in 1965.[39] In *The Young Offender* (England, 1967),[40] D. J. West takes less than one page from 300 to comment on some possibilities, and on the findings of the

Television Research Committee; and in their book, *Delinquent Behaviour* (USA, 1966),[41] Martin and Fitzpatrick take just over a page to dismiss the view that the media are significantly related to delinquency. Horton and Leslie, in *The Sociology of Social Problems* (USA, 1965),[42] refer to the effects of television when they discuss mass communication, but not in their discussion of the causes and treatment of crime and delinquency. Marshall Clinard, in *Sociology of Deviant Behaviour* (USA, 1963),[43] is a little more interested as we have already seen and, in addition to his comments on newspapers and crime, referred to earlier, he uses slightly over three of his 670 pages to refer to the statements of Frederick Wertham and to comment on one or two media and crime studies which we shall come to later. In view of what was said earlier about the nature of concern, Clinard's concluding comments on this matter are worth quoting:

> "In most cases the result of the preoccupation of the public with the effect on juvenile delinquency of television, motion pictures, radio and comic books is merely to release the feelings that something should be done. The deeper question of why juveniles are interested in this entertainment raises issues which adults often do not wish to face because of their own interests in similar material. Likewise, this problem is evidence of a reluctance on the part of the adult world to deal effectively with factors basic to it . . . The existence of gangs of delinquent boys is a more important and more difficult immediate problem than television, motion pictures, radio, or comic books, but few communities have the necessary vision to attack it. In dealing with social difficulties the public tends to take the easiest course."[44]

There is no need to go further with these illustrations for they are reasonably representative. As far as we can ascertain, most of the psychologists, sociologists and criminologists who have taken delinquency as their main topic of research or study do not appear to regard television or any of the other media as a major cause—or even as an important contributory factor to the development of delinquent behaviour. However, some sociologists, like Howard Becker, are interested in the relationship between society's reaction to deviance and the form and extent of deviance. The media may play a part in this connection.

This lack of an established relationship does not mean that the matter is closed. There are many other questions in this general problem area that remain unanswered. It could be that the fact that television is missing from the list of causative or contributory factors is more a function of the inadequacy of the relevant theories and corresponding research approaches than it is indicative of the real nature of things. So far, none of the available theories of crime causation, psychological or sociological, can be said to be entirely satisfactory, although—given the nature of the problem and the state of social science—this should surprise no-one. Looking for simple, single causes of delinquent behaviour is not likely to be a very profitable exercise, and gradually more attention is being given to the search for a deeper understanding of the conditions under which delinquent behaviour appears, develops and spreads.

The multi-disciplinary nature of the enquiries, and the existence of different schools of thought within each discipline, is bound to lead to disagreements in several areas. However, there are at least two points on which there appears to be fairly widespread agreement and they are, first, that no single factor can be assumed to be the cause of delinquency and, secondly, that the task of linking together the various contributory factors in any given case is likely to be a difficult one.

Fifteen years ago, when Marie Jahoda[45] and her colleagues were investigating the impact of literature, they came up against the same problems as we have encountered, and they too were unable to find many references to the influence of reading matter in the studies on juvenile delinquency which were then available. They did not regard the absence of such references as accidental, but saw them as reflecting a broad trend in the exploration of social pathology. At that time, it appeared that those working in this field, being fully aware of how difficult it was to discover the causes of human behaviour on a purely empirical level, were turning to psychological theories for help and guidance.

In psycho-analytical theory they found explanations which indicated that various forms of behaviour disorder could stem from a warped personality which had its roots in the experiences of early childhood. The real situation, the actual family-parental or other relationship, was seen to be more likely than any reading matter to provide the provocation or the stimulus for anti-social behaviour.

The passage of G. K. Chesterton from his essay *The Fear of the Film* is still worth quoting:

> "Long lists are being given of particular cases in which children have suffered in spirits or health from alleged horrors of the Kinema. One child is said to have . . . killed his father with a carving knife, through having seen a knife in a film. This may possibly have occurred; though if it did, anyone of common sense would prefer to have details about that particular child rather than

about that particular picture . . . Is it that the young should never see a story with a knife in it? It would be more practical that a child should never see a real carving knife, and still more practical that he should never see a real father."[46]

It was appreciated by Jahoda and her colleagues then, as it is by us now, that this represents an oversimplified view of this particular approach to the study of delinquency. We recognize that there are other psychological approaches. Suffice it to say that there is a fairly substantial body of opinion that accepts that some of the vital causes of delinquency must be sought in the early experience of the child. No one, neither fifteen years ago nor today, would want to leave the matter there, for even working from the simplest possible level it can be assumed that behaviour is a function of both personality and environment. Consequently, in her enquiry into the impact of literature, Jahoda tried to find out whether reading matter formed a significant part of the environmental conditions which, together with certain personality predispositions, might lead to delinquency.

Again the answer was negative, and again she felt that this finding could be related to a broader psychological principle, namely, the one that holds that direct experiences are likely to have much more impact than vicarious ones. Broken homes, inadequate parents, and peer group influences, were all cited as more likely to be influential than reading material, and were used to illustrate the primacy of direct over vicarious experiences—a phenomenon which incidentally is not confined to the sphere of delinquent behaviour.

The implication was not that vicarious experiences are necessarily weak. It was accepted that, given the appropriate predisposing factors, reading a book might have a trigger function and in some circumstances might precipitate delinquent behaviour. However, no evidence could be found to support even this trigger effect, and the conclusion reached after completing the enquiries along this line (the relationship between reading and anti-social behaviour) was that there was no evidence for the allegation that criminal acts were likely to be initiated by the reading of 'bad' books.

As we have seen, the position in which we find ourselves fifteen years later is not very different. Still, there have been some developments in psychology and particularly in sociology during this fifteen-year period, and we shall look at this briefly if only to see if they offer the possibilities for further development and enquiry.

Our intention is to pay more attention to the sociological approaches, although before doing this we must at least show that we recognize that the few brief references to psychological theories and findings that were made in the discussion on reading material by no means exhausts the approaches from what may be termed the individual as distinct from the environmental angle.

Nigel Walker offers what he calls 'individualistic explanations'[47] of delinquency in terms of (1) inborn differences between delinquents and non-delinquents, whether congenital or hereditary, (2) maladjustments resulting from faulty upbringing, (3) weak ethical learning as a normal result of working-class ways of child rearing. He distinguishes these types of explanation from environmental explanations in which he includes economic and sub-cultural factors.

Leaving aside the congenital and hereditary factors,* we are left with an individual-family set of explanations, and the question we have to ask in relation to these is not just whether the influence of television has been found to fit into this area so far, but whether there is any likelihood at all of it being found in this area should the appropriate enquiries be carried out.

To be more precise, we are saying that explanations of delinquency have been offered in terms of personality deficiencies, emotional disturbance, broken homes, maternal deprivation, paternal absence, lack of love and affection, general parental inadequacy, erratic training, bad example in the home, and so on. Martin and Fitzpatrick, in their book *Delinquent Behaviour*,[48] group together family-centred and individual-centred explanations, and consider that some of the approaches that link personality factors to family interaction have much to offer in explaining delinquency. However, this is not our main point, although perhaps it needs stressing in view of the attention that will be given later to sociological approaches. Our task is to enquire into the possibility of a relationship between any one of the above-mentioned factors and the use of television. We shall return to this question later when, amongst other things, we shall look at the concept of identification, and examine what is known about the use of television by maladjusted children.

In the meantime, it is worth noting that we have changed our ground a little. In reviewing the literature, we are no longer concentrating on the use of television as a simple or direct causal factor in delinquent behaviour. We are not asking the question: Does television cause delinquency? Rather are we attempting to identify

* In doing this, we are not denying the existence of some inherited factors which may contribute to some forms of criminal behaviour, particularly to that of serious and persistent criminals. However, these inherited factors are unlikely to play any significant part in the behaviour of the general run of juvenile delinquents who are central to our study.

the 'causes' of delinquent behaviour, and then trying to find out if these are, or could possibly be, related in any way to television usage. This is in keeping with the multi-factor approach mentioned earlier, and is consistent with the growing tendency to think not so much of direct causes, but more in terms of the factors associated with the appearance, development, facilitation, reinforcement, rationalization and diffusion of delinquent behaviour.

Nigel Walker, in writing about what he terms 'the factor-interaction assumption',[49] suggests that, no matter how one sub-divides delinquents, it will probably be found that no single factor is peculiar to them as distinct from non-delinquents. The thing to look for is the combination of two or more factors where there are perhaps moderate deviations from the norm. We might find, for example, when comparing delinquents and non-delinquents, that some of the heaviest television viewers and some of the worst cases of maternal deprivation were in the non-delinquent group. The distinguishing feature between the two groups could be in terms of a particular combination of both factors within the same family. However, even in these cases, it is not always possible to say anything very definite about the consequences or actual outcome of such a combination.*

We do not see it as our task in this book to argue the pros and cons of the several different approaches to the study of delinquency, but a few words of caution are necessary against a too ready and uncritical acceptance of what on first sight appears to be an attractive and useful method.

The multi-factor approach† has been extensively used in the study of delinquency, probably because its general eclecticism enables it to cut across many different levels or kinds of theory. The main idea is to compare delinquents and non-delinquents on a number (often a large number) of items in an attempt to identify those items which are associated with delinquency. The higher the degree of association, the more importance is assigned to that factor as a 'cause' of delinquency.

One of the main difficulties in this approach stems from the over-simplified, relatively crude, dichotomous classification system which is frequently used. For example, the delinquent category—not surprisingly—usually consists of those who have been caught.‡ Unfor-

* It is appreciated that all these approaches are sometimes referred to as causal studies.

† The discussion of this method draws heavily from Martin and Fitzpatrick's *Delinquent Behaviour*.

‡ Self-reporting offers a way out of this difficulty but presents several others, as we shall see.

tunately from the research point of view, delinquency statistics do not define the delinquent population. Consequently, the chief differences between the official or formal delinquents and the non-delinquent may tell us more about people who get caught (this need not be unimportant) than about the causes of delinquent behaviour. The whole process of differential detection, law enforcement and sentencing is relevant in this connection. In these studies, a great deal also depends on the particular stage in the delinquency-punishment process at which the data are collected. The effects of detection, the influence of the police, the nature of any institutional experiences, and the general impact of the judicial process, cannot be discounted.

Another problem is that the approach seems to facilitate the explaining away of one bad thing by another bad thing. On the whole, it does not allow for several 'bad' things (e.g. delinquency, broken homes, cognitive poverty) to be regarded as related parts of a more general condition which may not be regarded as 'bad' (e.g. our socio-economic system). Social problems can be seen both as the price we have to pay for the particular set of institutional arrangements we have 'chosen', and as the indirect, unanticipated consequences of otherwise 'good' actions.

When one reviews the many approaches to the study of the causes of delinquency, it is easy to see why some students have found it necessary to reject the idea that any single theory could possibly be adequate for their purposes. Quite a few have followed such a rejection with an acceptance of the multiple-factor approach but, in doing this, some of them appear to have confused explanation by means of a single theory with explanation by means of a single factor. A single factor tells us little or nothing. Ideally, a single theory is capable of presenting *an* explanation of how several factors are linked together to explain delinquency.

Factors should not be confused with causes. Factors have to be treated as variables and linked together in some logical and meaningful way before they can be used to explain a process that leads to delinquency. When this has been accomplished, we might then have a theory of delinquency causation. As we have noted, there are in existence several such theories, each one emphasizing the causal importance of a certain set of variables. Whether these theories overlap or are in conflict with each other depends to a large degree on the limits they set themselves. Obviously, a theory which deals with juvenile gang behaviour need not be in conflict with one that attempts to explain sexual deviance.

This point on the limitations of specific theories should be borne in mind during the brief discussion which follows on environmental or sociological approaches to the study of delinquency. There are many such approaches which range from relatively crude explanations in terms of absolute poverty, through relative deprivation and anomie, to more sophisticated ones in terms of sub-culture and drift. Not forgetting that our main concern is with the relationship between television and delinquency, we must survey—albeit briefly—what we consider to have been the most important attempts to unravel the delinquency problem. Let us remember that we are looking not only for direct references to the mass media but for any indications of the role that television or the other media might possibly play in the development of delinquent behaviour. Once again, this time from the sociologists, we get a more or less negative response to our first question, although Sutherland and Cressey[50] do have a reference to the press and the glamourization of criminals. We shall see if we get much further on the second question.

The main distinction between what have been called the individualistic or family-centred explanations on the one hand, and environmental or societal explanations on the other, is that the former deal with causes in the individual or in his upbringing, whereas the latter emphasize the part played by economic, ecological or social conditions.* More specifically, the mainly sociological approaches which are discussed below tend to regard delinquency as a social phenomenon, a style of social behaviour, something that is learned in social-interaction. It should also be remembered that the emphasis in sociological theory is not on explanations of individual delinquent behaviour, but on the history, definition, rationalization, organization and transmission of delinquency in society.

Our starting point is that "criminal behaviour is *behaviour* and, like other social behaviour, criminal behaviour must be learned".[51] From this we may proceed to the theory of differential-association which is based on learning and was fathered by Edwin Sutherland, and developed by Donald Cressey.† "The hypothesis of differential-association is that criminal behaviour is learned in association with those who define such behaviour favourably and in isolation from those who define it unfavourably and that a person in an appropriate situation engages in such criminal behaviour if, and only if, the

* There is clearly a degree of artificiality in these distinctions.

† The theory was also modified by Daniel Glaser[52] to 'differential identification'.

weight of the favourable definitions exceeds the weight of the unfavourable definitions."[53]

Although one would not think so from the confusing controversies which have raged around it, this represents a fairly simple and straightforward position. In addition to maintaining that delinquent behaviour, like any other social behaviour, is learned but that the learning should be thought of in sub-cultural rather than in individual learning terms, the central features of the position are, first, that the world of criminal or delinquent learning should not be regarded as a separate world, for all of us in one way or another and in varying degrees have associations of both a delinquent and non-delinquent nature. Whether an individual is law-abiding or not, together with his attitudes towards the law, should be seen as a function of the nature, frequency, duration and perceived prestige of his various associations. The definition of the situation by the delinquent or groups of delinquents, and their attitudes about the 'legality' of their motives and behaviour, are also central to the general position. The media are seen as working in association with personal factors to influence receptivity to criminal behaviour and to produce incidental offences.

Walker probably under-estimates the value of this approach when, in accepting its limited usefulness in explaining the acquisition of techniques and in accounting for the *precise form* which the delinquency takes, he writes: "we should not altogether dismiss differential-association as a sterile hypothesis".[54] It obviously has its limitations (although it has often been criticized for failing to do what it never set out to do), and a particular criticism has been that it has not provided accurate specifications of the learning process. Perhaps its greatest contribution has been to provide principles or a framework for organizing the relevant data. The need for this has already been noted.

Other criminologists have also given emphasis to the factors which are at the core of differential-association, i.e. the social nature of crime, learning through reference groups, and the individual's definition of the situation. The concept of 'reference idol' has also been used and this could suggest a part that might possibly be played by the mass media in this learning process. In addition, the concept of differential identification — and the importance attached by some sociologists to the concept of self-image in the development of delinquent behaviour—suggest other 'media possibilities'. The media may also play a part in assisting the individual to define his situation and in the formation of attitudes towards the law. Moreover, the theory allows for law-violation being learned through intensive association or identification with other offenders as they may be *both psychologically and actually available* in the operating milieu of given individuals.

David Matza[55] writes of the drift to delinquency, a drift which he feels is often aided, albeit unknowingly, by society and the custodians of law and order. Earlier, in co-operation with Gresham Sykes,[56] he had suggested that when behaviour deviates from the ideal it may be motivated by 'subterranean values'. These are the values which, although at cross-purposes with other deeply held values, may nevertheless be widely recognized and accepted in society. The legal order may be violated by those who subscribe to it. These values may or may not lead to overt delinquency. Doing things for 'kicks' and a contempt for work are examples of such values.

At the same time, so as to neutralize the 'moral bond to law', to allay the guilt which may follow when internalized norms are violated, 'techniques of neutralization' are developed. These include the denial of responsibility, the denial of intent, the denigration of the victim's status, the questioning of the moral status of control agencies, and reference to more important loyalties of the peer group. Using this two-barrelled approach (subterranean values and techniques of neutralization), it can be argued that rather than being opposed to the adult world, the young delinquents are in fact reflecting it in their delinquent behaviour. The breaking of the law by delinquents can be seen as the result of relationships with an inconsistent and vulnerable legal system.

These neutralizing techniques which the individual may learn, probably in social conditions which spawn perceived injustice, can set the young person free from his allegiance to 'right and proper' behaviour. Once free in this way, the young person *may* drift towards delinquency. Different degrees of neutralization may be reflected in different degrees of delinquent participation. As the techniques are learned this gives this approach something in common with differential-association, although this latter approach emphasizes the difference between delinquent and conventional values, whereas the former stresses the similarity between them.[57] It seems possible that the media could play a part in facilitating the learning of some of these techniques. The questioning of the authority and status of control agencies immediately springs to mind.

However, even if the young person has started to drift and is no longer subject to social control, he still need not indulge in delinquent behaviour. The motivational push may be lacking. The concept of 'will' is used by Matza to cover this point, and it is argued that the will is likely to become operative when two conditions are present: that is, when both the skills and fears can be successfully managed by the adolescent, and when 'desperation' develops as a result of his 'being pushed around'.

This kind of motivational problem has been dealt with in a somewhat simpler way by Irving Targoff,[58] who looks at delinquent behaviour in terms of the interplay between temptation and control. He regards delinquency as a function of the balance between the two forces in any given situation.

It is also possible to see a link with the differential-association approach in the work of some of the interaction theorists, notably Howard Becker and Erving Goffman. The approaches have several features in common, perhaps the most obvious being the stress on the delinquent's definition of the situation. The emphasis in the interaction approach is on delinquency as a social role, developed for each individual through a career of social interaction.

Although we cannot credit R. K. Merton[59] with the origination of the concept of anomie, it is with him that we shall start our brief discussion of delinquent sub-cultures. This is largely because his theories have been the starting point for important work in this area over the last 10 to 15 years by such people as Albert Cohen,[60] Richard Cloward and Lloyd Ohlin.[61]

Merton looked at the imbalance or disjunction between the officially approved goals of society (in the USA) on the one hand, and the legitimate or approved means of achieving these goals on the other. The goals held up before everybody were those of material success and prosperity, but not everyone had a chance to use the approved means of reaching the goals. There was no equality of access to legitimate channels. Poor people had their hopes and aspirations stimulated just like their more fortunate brethren. The pressures were there all the time, and there was a great temptation to take the short cut to success. Thus we have the illegitimate or delinquent solution.

'Innovation', the name given by Merton to this form of adjustment, which involves an acceptance of approved goals and a use of illegitimate means, is but one of five ways of responding to the anomic situation. The others are 'retreatism', which involves contracting out and repudiating or rejecting both ends and means; 'ritualism', which involves acceptance of—in fact obsession with—the institutionalized means, the goals having become unattainable; 'conformity', where both means and ends can be accepted; and 'rebellion', where change is sought in the goals and possibly also in the institutionalized means of attaining them.

This is the basic position from which many others have started in their attempts to study those groups in our society which deviate from or positively reject the morality of the majority. Albert Cohen suggested that the behaviour of delinquent gangs stems from the basic hostility of the gang members towards middle-class values. (The working-class boy, lacking status, resents the dominant values—ambition, self-reliance, respect for property etc.—of a world in which he plays no part.) They also resent the people of this middle-class world, e.g. teachers and police, whom they regard as wanting to devalue the working-class youngsters, just because these youngsters lack the prized middle-class virtues. This makes the situation worse and increases the hostility.

Cohen argues that a working-class delinquent gang is a natural outcome of this situation. The young males with common hostilities get together both to achieve status and to get their own back on those they believe to have offended them. In addition to its opposition to middle-class values, the sub-culture* tends to be characterized by opposition to all things virtuous, a desire to annoy and irritate the representatives of respectability, short-run pleasure seeking, versatility in its delinquent activity, and by group autonomy and opposition to external control. Status is gained within this framework of values and activity.

There have been several criticisms and some developments of Cohen's position which we can only touch on here. The main criticisms have centred on the assumption that there is a sharp break between the values of the middle and working classes, that the working class as presented by Cohen is much more homogeneous than one has any right to believe, and that no adequate explanation is offered as to why the rejection of the middle-class position should take the *specific form* that it is assumed to take in this approach.

Cloward and Ohlin's theory, which is more dependent than Cohen's work on Merton's scheme of socially structured anomie, is nevertheless closely related to this work in that it also puts forward the idea of socially induced tension states based on inter-class conflict: "The disparity between what lower class youth are led to want and what is actually available to them is the source of a major problem of adjustment. Adolescents who form delinquent sub-

* There is some confusion in the use of this term. Sub-culture is used in several ways most frequently with reference to the norms, values, beliefs etc. of groups within a society. Contra-culture involves value conflict, and counter-norms, which stem from strain between a sub-group and society as a whole. Cohen's theory is seen by some as spelling out a contra-culture.

cultures, we suggest, have internalized an emphasis upon conventional goals. Faced with limitations on legitimate avenues of access to these goals, and unable to revise their aspirations downwards, they experience intense frustrations; the exploration of non-conformist alternatives may be the result."[62]

The frustrated individual who blames society for his predicament is likely to join the particular anti-social group which is most 'suitable', or the most readily available in his area or neighbourhood. Whether he joins the 'criminal', 'conflict', or 'retreatist' sub-culture depends on opportunity and status possibilities. Apart from any other function they may perform it seems likely that the media could play a part in conferring status on certain forms of group membership and behaviour.

There are many criticisms that can be and in fact have been made about this approach. We have already come across some of these, at least implicitly, in the work of Matza who tends to think of the young delinquent evading rather than opposing the dominant morality. One would also like to know more about the level of aspirations both within and between the social classes, for it surely cannot be as stable and uniform as it appears to be from this work. We must also ask whether the avenues to success and the legitimate opportunities are as firmly blocked for these young people as they appear to be. This is a question that certainly has to be asked in this country. These and other questions and criticisms can be made without invalidating the basic idea of a delinquent sub-culture, but we should be on our guard lest we go too far in our acceptance of an all-powerful, anti-establishment mentality. As West has written: "In England with its more muted versions of social conflict the delinquent allegiances are likely to be more mixed and vacillating."[63]

These explanations and speculations can appear extremely attractive and, notwithstanding their American base and the fact that they have often been loosely applied in this country, they are clearly not without relevance for us. In fact, the work as a whole has much to offer both with regard to the learning of delinquent behaviour, and to increasing our understanding of the way society produces the conditions in which delinquency develops. Nevertheless, questions of national application apart, a further word of caution is necessary. As Walker reminds us, "the intellectual attraction of a theory must not be confused with the evidence for it . . . so many associations of different sorts are recorded in the literature that it is not difficult to assemble a small collection to fit almost any intellectually plausible explanation".[64] West, too, draws attention to the "outstanding weakness of the social theories of delinquency", due to the absence of factual evidence in support of any one in preference to the rest.[65]

For those who wish to delve into the intricacies and speculations of the sub-cultural (and contra-cultural) debate, one can do no better than recommend that they read David Downes' book, *The Delinquent Solution*.[66] He recognizes and discusses the weaknesses in the sub-cultural approach without being able to avoid some of them in his own work. This work also brings us nearer home, and has the advantage of reporting research carried out in this country.

Downes also reminds us that sub-cultural theorization is not the same as validation, and he maintains that the theory is likely to survive only if it proves susceptible to considerable modification. He argues that its limitations must be recognized, particularly with regard to its application in this country "where delinquency in general is hardly a major social problem, where gang delinquency on the American model is non-existent, and where lack of research makes for difficulties not experienced by American investigators".[67] It is worth noting that this work was published in 1965, and it could be that, at least in some respects, the situation has changed since then.

As early as September 1966, N. Howard Avison was arguing that an organized criminal class was emerging in this country and that we could expect juvenile gangs to play a more prominent part in the English delinquency scene within the following five years.[68]

Downes feels that what evidence there is points to the existence of a sub-culture, rather than a contra-culture, in the large urban and metropolitan areas in this country. He considers that the real problem is to ascertain whether or not, or to what degree, the male working-class 'failure' is faced with a problem of adjustment at all.

The answer Downes gives to this question from his own research is that the English 'corner-boy'* probably does have a problem of adjustment, but it is not the same as the one experienced by his American counterpart as indicated by Cohen. Although engaging in delinquent behaviour (on the whole petty) the expressed norms, values and beliefs of the young offenders studied by Downes in London scarcely differed from those of the adult lower working class, and were essentially conservative. Downes suggests that the illegal behaviour seems to be due not so much to 'alienation' or status frustration, as to a process of *dissociation* from the middle-class dominated

* The 'corner-boy's' life is organized round working-class values and, although middle-class values would be regarded as interfering with adherence to the 'corner-boy' culture, this culture is not specifically delinquent.

contexts of school, work and recreation. This leads to an over-emphasis on purely leisure goals "sedulously fostered by commercial teenage culture", rather than on other non-work areas. The commercial teenage culture is seen as filling the gap created by the dissociation from school and work. Aspirations are generated which cannot be fulfilled and the delinquency springs from the situation where the individual needs to achieve his leisure goals, but where it is impossible or extremely difficult for him to achieve them by legitimate means. That the mass media *could* play a part in this process seems beyond doubt.

For the most part, the reaction to 'failure' is not frustration and repudiation, as suggested by the American theorists, but the reaffirmation of the working-class value system with its traditional constraints. This is akin to Cohen's "revising his aspirations downwards". However, it is suggested by Downes that in one sphere, that of leisure proper, the old values have lost their hold. In this sphere, the traditional solutions no longer apply for the 'corner-boy' with 'time on his hands' and 'nothing to lose'. The delinquent sub-culture offers the way out, providing a solution to an intensely felt problem. Working-class non-delinquents and middle-class young people may also face problems of adjustment, but Downes' point is that the working-class 'corner-boy' is the only one to depend exclusively on leisure as the framework for 'exploit'. It is also emphasized that any general increase in 'dissociation' from the realms of school and work, and exacerbation of leisure goals as a consequence, will lead to more of this type of delinquency.* The source of the working-class male adolescent's leisure problems originates in the school and work situations to which he has been allocated.

Jackson Toby[69] also stresses the importance of success or failure in the educational process. For children who fail to make the grade, school can become the place where the battle against society begins. According to Toby, relative to adults, the adolescent feels deprived as a member of a powerless minority group. The school failure is likely to have these feelings intensified, particularly if—as is quite likely—he is also conscious of his relatively deprived economic position. The mass media, so it is claimed, play a part in this explanation, for they are seen to stimulate the desire for a luxurious type of life and to be furthering the development towards a consumption-oriented society. The young working-class failure is considered to be

* There are other patterns of delinquency besides the one described here, but this pattern probably accounts for a serious proportion of group juvenile delinquency.

particularly vulnerable to these commercial pressures of the affluent society, although the precise nature of the link-up of all these factors is not specified.

Bryan Wilson,[70] a sociologist, but—like Toby—not thought of primarily as having his main research interest in this field, suggests that social change and unfulfilled economic aspirations could be at the root of delinquent behaviour and vandalism. A central place is given in his argument to the effect of the induced achievement-orientation of contemporary society, for not all who are invited and pressured to get on actually want to get on. A widely diffused sense of frustration and discontent is the result. Wilson states that the limitation of opportunity is not just a matter of persisting inequalities between classes, it is also the relation of reality to the over-stimulation of aspiration.

In this account the mass media are mentioned from time to time. Young persons are said to react against the increasing impersonalization of modern life, and to the breakdown of the nexus between home and school, by having greater recourse to the values of 'the entertainment-dominated youth culture'. The success values of superiority are widely disseminated by the media. Star entertainers are seen as 'inverted scapegoat symbols'. They represent success without effort or virtue. These people do not simply facilitate sublimation, they further frustration by exaggerating and exacerbating the success ideal, and by failing to give ethical emphasis to the appropriate and institutionalized means of achieving success. The media also advertise the means whereby real 'kicks' are to be had, the need for which stems from the boredom and frustration of living in an impersonalized, over-institutionalized and routinized society.

It is suggested that youth culture, which is neither confined to the working-class nor necessarily delinquent, by its emphasis on protest, 'kicks', and short-term satisfaction, erodes moral restraints and lowers resistance to delinquent temptations.[71] There is not a great deal that is new here and, once again, particularly with reference to the influence of the media, it is very much a matter of the predominance of speculation (interesting and stimulating though it is) over facts.

There are ways other than inter-class tension of looking at gang delinquency. For example, Bloch and Niederhoffer[72] view the gang not as a phenomenon rooted in inter-class tensions and/or blocked opportunities, but simply as something that develops to meet adolescent needs. According to them, delinquency is widely diffused throughout the class structure.

Short and Strodbeck,[73] in an important and interesting study, suggest that the important goals of the wider society do not seem to be the main source of frustration for the gang boys in their study. Gangs are not as prominent in England as they are in the United States, but the finding that for some adolescents it is their lack of training and their inability to handle a variety of social situations and deal with changes that lead them to develop a youth culture, is none the less an interesting one and may apply in other circumstances. The youth culture, with its stress on stylistic distinctions and its concern with immediate gratification, meets their immediate needs and makes it possible for them to avoid exposing their incompetence in interpersonal relationships.

Walter Miller[74] suggests that some of the values which are characteristic of lower-class gang delinquency, values which reflect the adult, lower-class world ('fate', 'toughness', 'smartness', desire for excitement, and status needs), are gradually being spread to the middle classes by the mass media. At the same time, Miller sees a built-in conflict between the life styles of the lower class and the middle class. Much of this is in the realm of speculation, and there are of course several other ways of accounting for the spread of middle-class delinquency which need not concern us here. One must admit that, in reviewing these various sociological approaches, from time to time one is reminded of Nigel Walker's cautionary, almost despairing, concluding comment in his review of environmental theories: "Until sociologists have accepted these standards of evidence and produced studies which comply with them, it might be thought a waste of time to carry the discussion of causal factors further."[75]

We shall not carry this particular discussion any further, although our reasons for not doing so are somewhat different from Walker's. We have not attempted a comprehensive survey of all the work in the field, but we feel that we have covered enough ground and spread our coverage in such a way that we have a reasonably good idea of what those who have studied delinquent behaviour have to say about the links between that behaviour on the one hand, and the mass media in general and television in particular on the other. As far as hard, reliable evidence is concerned, it would appear that the situation has not undergone any great change since Frederick Thrasher[76] reported on his study of the relationship between comic books and delinquent behaviour in 1949. Martin and Fitzpatrick[77] echoed the findings in Thrasher's report when, fifteen years later, they concluded that there was "no acceptable evidence that television, movies, pornographic literature, and other forms of mass media, are significantly related to delinquency". They also suggested that the concern reflects "a search for meaning in an ambiguous situation", as perplexed, troubled and concerned people try to make quick sense out of the many confusing social problems and rapidly changing social conditions that surround them.

We made a similar point earlier in this book when examining the nature of concern, so obviously we are not out of sympathy with the view just expressed. On the other hand, it needs to be stressed that all that has been established so far is that we have been unable to find in the work of those who specialize in the study of delinquent behaviour any acceptable research results which would enable us to conclude that television may be regarded as an important causal factor in delinquent behaviour.

However, we are unable to close the book yet, and this for three main reasons. First, because we certainly cannot assume that all the relevant questions have been asked by those who study delinquent behaviour. The general state of the field, as we have seen, suggests that it would be unwise to assume otherwise. Secondly, because there is another field of enquiry, namely mass communication research, which we have yet to review, and thirdly, because even though many of what might be termed indirect references to the mass media in the literature just reviewed tend to be rather speculative, we are not at this stage justified in dismissing them out of hand. There are leads or hypotheses here that demand further investigation.

Several years ago, Gresham Sykes[78] emphasized that rather than "looking for devils in the mind or stigmas of the body", our search for the factors which might be linked to delinquent behaviour should be in the *ever-changing relationship* between the individual and the social group to which he belongs. Today we would follow a similar line and draw attention to the need to study the delinquent not as an isolated individual who is, or became, a delinquent at some given point, but as someone who is interacting all the time with other people, in a variety of situations where different roles may be available, and changes are always possible.

It is because of this approach that at least some of the delinquency–television options must be kept open. In fact some new ones might be created. For example, if delinquent behaviour is learned, and if the learning takes place primarily in inter-personal situations, would it be more profitable to concentrate not, as often in the past, on direct learning from the media—but on the way the media provide general perspectives towards negal norms?

Leslie Wilkins[79] provides us with another possibility when he suggests that our knowledge and particularly our personal experiences of behavioural variations in society could lead to greater tolerance. We might ask what is the nature and extent of our knowledge about delinquent and other forms of deviant behaviour in our highly organized 'protective' society. In our society the deviance that is revealed—perhaps still more the deviance that is reported—is given the appearance of being more unusual and extreme than it really is. This could lead to corresponding extreme reactions on the part of the general public and gradually a wider range of both behaviours and people may come to be regarded as deviant in some way or other. Finding themselves so labelled or stigmatized, these people turn inwards and develop their own values. This leads to still further rejections and to still greater emphases on conformity by the community in general. It is a vicious circle based on inadequate or inaccurate information. The role of the media in amplifying problems, particularly with regard to new social phenomena (e.g. skinheads) demands our attention, but first we must examine some of the more conventional approaches to the study of the influence of television.

In singling out for separate attention mass communication research, or at least that part of it which deals more directly with the media–delinquency relationship, it is appreciated that the boundary lines must be somewhat arbitrarily drawn and that there are bound to be considerable overlaps both with the work just covered, and in particular with mass communication research of a more general nature which will be covered later. Still, it is possible to identify a considerable body of literature which in some way or other has dealt with the media–delinquency problem. The literature records a variety of methods and approaches in addition to various forms of the type of study reported in detail in the second part of this book. These include analyses of media content, investigations of the media activities of delinquents, the questioning of both delinquents and others—such as psychiatrists and social workers—on what they felt about the contribution of the media to the delinquency, and laboratory or experimental studies on the media portrayal of violence and aggression.

We saw earlier that, in many ways, what people are now saying about the effects of television they have said before about the effects of the cinema, comics, magazines and books. As each medium produces its own wave of concern, so also does it lead to its own set of enquiries. We have just read that Frederick Thrasher reported on his work on comics and delinquency in 1949, and John R. Cavanagh[80] neatly summarized the position around that time when he stated that no-one had conclusively demonstrated that comic books were detrimental in any way, and that no normal child under the age of 12 was likely to be harmed by them. He argued that neurotic children needed treatment and would be equally affected by the movies or the radio, and that campaigns to eliminate comics were useless, serving only to release the aggressive feelings of the crusader.

Five years later, Marie Jahoda,[81] in the study on the impact of literature already mentioned, made the point that there was a large overlap in content matter between all the media, and that it was virtually impossible to isolate the impact of any one of these on people who are exposed to all of them. The unlikelihood of attitudes being changed as a result of exposure to the media was also stressed, although it was accepted that young children, who have not yet crystallized their attitudes, may be particularly open to influence in this respect. The vulnerability of the insecure or otherwise maladjusted child was also recognized, and it was thought possible that compensatory viewing or reading by such children could lead to a reinforcement or intensification of the problem and to a greater readiness on their part to accept violence and brutality in the real world.*

One of the earliest studies on the influence of cinema on delinquency was reported by Herbert Blumer and Philip Hauser[82] in one of the Payne Fund studies of the early 1930s. After carrying out several forms of enquiry, Blumer and Hauser found that about 10 per cent of the delinquent boys and 25 per cent of the delinquent girls in their studies had been influenced by the cinema in some way. It was argued that on the whole the influence would probably be unconscious and indirect, and would be likely to stem chiefly from films which portrayed criminals, created a tough mentality, induced sexual desires and emphasized wealth and luxury as well as the undesirable ways of achieving these.

'Direct' influences were also mentioned and these included the spreading of knowledge about criminal techniques. It was accepted that the young person's repertoire of deviant skills could be increased, although it was appreciated that the anti-social ideas provided by the cinema could be entertained without these necessarily leading to overt anti-social behaviour. It was emphasized, however, that the skills and ideas were there for later use should conditions be appropriate.

* In a concluding note, Jahoda attempts to give much-needed perspective to this type of enquiry by asserting that the civilizing effects of literary production and of the other mass media of communications present one of the few unquestionable achievements of our time.

The importance of general social conditions was fully recognized by Blumer and Hauser who suggested that the influence of the cinema was probably proportionate to the weakness of family, church, school and neighbourhood. In drawing attention to the possibility that films of the 'crime does not pay' variety could have beneficial effects, a cautionary note was sounded, for it was appreciated that—no matter what the intention—such films may not have the desired effect. The factors which it was thought might weaken the beneficial influence of such films included the sympathy for the criminal which was often aroused, as well as the glamour frequently associated with the delinquent role and the attractiveness of the promise of easy gain and adventure.

In the same year, reporting on research which had been supported by the same fund, Paul Cressey and Frederick Thrasher[83] suggested that a correlation existed between truancy and delinquency and frequent attendance at the cinema. Whilst accepting that it would be wrong to state that those who attended the cinema frequently were not influenced by the cinema, they felt that it would be equally wrong to hold that films were solely responsible for anti-social behaviour and delinquency. In this work they came face to face with the problems which correlational studies often bring in their wake. In this case, does excessive reading, viewing, or cinema attendance lead to delinquency? Do delinquents tend to be more frequent users of the media, or is it that both delinquency and heavy media consumption are related aspects of a wider condition?

A survey carried out in West Bromwich in 1950[84] led the researchers to report that, despite the popularity of crime films, they could find no relationship between cinema attendance and juvenile delinquency. The danger of the recurrent portrayal of idealized luxury was again singled out as a *potential* danger.

Although most of the work normally referred to in this field is reported in the English language and has been carried out in Anglo-Saxon countries, it would be a mistake to ignore important work on the cinema from France, Italy, West Germany and Sweden. This work cannot be covered in detail here and, in any case, up-to-date summaries are available.[85] However, it is worth noting that Monfredini,[86] commenting on Italian research, refers to work that indicates that social scientific research on the subject of the cinema as a cause of delinquency is likely to encounter practical obstacles, perhaps even insuperable ones, because of the complexity of the situation, the interplay of numerous variables, the difficulties of defining the variables, and the problems of establishing adequate controls. The general conclusion from the Italian work seems to be that whilst the cinema can work to the detriment of some people in some situations, it cannot be said to be a main cause of delinquency, for delinquency is not susceptible to such simple analysis.

Summarizing the results of the most important research carried out between 1929 and 1956, Decaigny[87] stressed that the cinema was not the only leisure-time activity of children and that leisure-time activities were not the sole cause of delinquency. A film, so it was argued, may suggest the *modus operandi* but it is never likely to be a primary cause. The fact that no two children are likely to be influenced by films in the same way or to the same extent, and the overall tendency of the media to reinforce what is already there, rather than form new ideas or bring about fundamental changes, was also stressed in this work.

According to Decaigny, the 'emotional effect' of a film on a child depends on the intensity of his involvement in the situation presented by the film, his capacity for reaction and criticism, and the degree of his confidence in the world and people around him. Also mentioned is the tendency of the cinema to present and spread a distorted conception of life.

An analysis by Clostermann and Preuss[88] of the case histories of 342 juvenile delinquents in Germany led them to conclude that film was not a primary factor in delinquency. The research covered by Furhammar,[89] in his review of work which had been reported in Scandinavian and German languages, tended to reinforce this conclusion.

Many studies have been carried out on the influence of the cinema and, in selecting from these as we have been doing over the past few pages, it is inevitable that some values or principles of selection will be employed. In reviewing the work on the influence of the cinema on children and young people, one cannot help but notice the multiplicity of approaches and the conflicting nature of the results. The disagreement or confusion is particularly noticeable when one looks at the many *interpretations* and *opinions* about the alleged relationship between the cinema and delinquency. As we shall see in greater detail when we discuss television, it is possible to interpret the same basic data in different ways.

In fact, one of the most interesting aspects of the whole situation is the way in which data are interpreted. It might be thought that the situation would become clearer if we could make a distinction between the results which derive from what may be termed acceptable social scientific research and those which stem from more impressionistic work. It would be difficult enough to make this distinction and more difficult still to obtain a reasonable measure of agreement on any distinction made, but unfortunately even then there would be no guarantee that the situation would be clearer. We are reminded in the UNESCO paper that "even the increasing availability and use of scientific research techniques has not noticeably clarified the situation".[90]

When all this is borne in mind, it is easy to see that the author of the UNESCO annotated bibliography on the influence of the cinema on children and adolescents had no alternative but to conclude that "on the evidence so far available, it is extremely difficult—indeed virtually impossible — to establish that the cinema has a direct influence on juvenile delinquency".[91] It is important to note that the conclusion was not that the cinema had no influence whatsoever, but that what influence it had was generally thought to be indirect, unconscious and probably effective only after a long period of exposure to repetitive themes and values. "The broad generalization might be made that the film has mainly a provocative effect but is rarely basically causal."[92]

Gradually the research effort shifted from cinema to television and three years after the publication of the bibliography on the influence of the cinema, UNESCO published another bibliography, this time on the effects of television on children and adolescents.[93] Quite a number of the studies referred to in the cinema publication and discussed above were also included in the television publication. Some new experimental work on the effects of the media portrayal of violence and aggression were reported in this second publication but there were few references to work which had attempted to unravel the delinquency–television problem. As we go on to examine the research on the effects of television, we shall see that the emphasis gradually shifts to the question of violence. The wider and, some would argue, more relevant problems have received relatively little attention in more recent years.

Research by the Australian, R. J. Thomson,[94] on the reactions of some young teenagers to two crime dramas, is worth mentioning in that Thomson, although failing to find any direct provocative effect with regard to delinquent behaviour, felt that there was a risk that constant viewing of crime dramas and related material could produce stereotyped reactions and lead to insensitivity. Of the large-scale general studies on television and children, Maletzke[95] in Germany could find no conclusive evidence that television caused juvenile delinquency, Himmelweit[96] in England had little to say about the matter as such, neither had Furu[97] in Japan. In the United States, Schramm, Lyle and Parker[98] concluded from their research that television could contribute to delinquency but could hardly be a basic cause.

Quite a number of studies were reported in the fifties and early 1960s, which illustrated in various ways that television could be used in different ways by different types of children. With television, as with the other media, the friendless, lonely, disturbed, unstable and maladjusted child (variously defined) was said to be particularly vulnerable or susceptible to the influence of the medium. Particular attention was given to the concept of identification, and it was argued that the degree to which a child identified with a television personality was probably crucial in determining the influence of that particular television experience.

We shall have more to say later about the concept of identification, but we can say now that, as yet, there is little support for the case that identification with television personalities is a major causal factor in delinquent behaviour. As the UNESCO review of the television–delinquency question almost predictably concludes, "The roots of this criminal behaviour lie far deeper than television; they reach into the personality, the family experience, the peer group relationships of the delinquent or criminal individual. At most, television can be merely a contributory cause and is likely to affect only the child who is already maladjusted and delinquency-prone."[99] It was accepted in this review that television might possibly play a part in triggering off delinquency, spreading information about delinquent skills and conveying distorted ideas about how certain goals might be attained. But it was stressed that "in any of these cases, television by itself cannot make a normal, well-adjusted child into a delinquent."[100]

A more comprehensive and thorough study on media and delinquency than many of those "based on experiment, survey or clinical study" which are reported in the UNESCO bibliography, was carried out by Edwin Pfuhl Jnr.[101] in partial fulfilment of a Ph.D. in sociology at Washington State University in 1960. Pfuhl, although conscious of some of the limitations in his work, nevertheless feels able to state that his data throw considerable doubt on the ideas that the media are an inducement to criminal behaviour, or that they facilitate the adoption of delinquent skills or techniques. Moreover, he found that his delinquents and non-delinquents did not show any clear differences in their interest in or exposure to crime themes in the mass media.

This particular study, largely because of its methodological relevance to our own research, is reviewed in more detail in the second part of this book. It is suggested there that had Pfuhl used different categories and different forms of analysis in his work he might have found that in certain more specific areas differences did in fact exist between his delinquent and non-delinquent subjects. This particular point will not be pursued further here, but in fairness to Pfuhl, it ought to be mentioned that it is by no means certain that all the research results reported in the various bibliographies, as well as those which are to be found in widely read authoritative summaries or in the evidence that is put before national commissions and the like, have been looked at as we later look at Pfuhl's work.*

Perhaps it is not necessary to argue that all work so used should be thoroughly examined. However, as we saw earlier, we would be unduly optimistic if we thought that by applying some rigorous set of rules we could easily obtain a clearer, more definitive and more widely agreed description of the effects of television than the one which is normally placed before us.

At a very crude and simple level we have used one form of selection and categorization in this book. This has placed a psychiatrist, Frederick Wertham, on the 'concern' side of the fence and not with the sociologists and psychologists. Wertham would probably welcome this distinction, although his reasons for so doing would probably not be the same as those implicit in our decision. In his article 'The Scientific Study of Mass Media Effects'[103] published in the American Journal of Psychiatry, Wertham shows little respect for the work of Himmelweit, Schramm and their colleagues, referred to earlier, and often regarded as classics in this field. Klapper, too, comes in for severe criticism. Wertham claims that in the questionnaire or survey approach the whole child is not examined and the essence of the television–child relationship is not touched. He writes of the "dubious scientific value" of such work, and argues that statistical relevance is confused with social relevance. He also suggests that his own clinical approach which he has applied to "over 200 unselected cases" is the only one that can provide "a truly scientific study of media effects".

This is not an issue that we propose to explore to the full, but neither is it one that can be dismissed quite so lightly as some of us have tended to do in the past. It is not just a matter of psychiatrists

* We are reminded that some years ago Chapanis and Chapanis,[102] after thoroughly reviewing the experimental evidence which had been put forward in support of dissonance theory, found numerous shortcomings which led them to conclude that the work would not stand up to close scrutiny and that the theory had not been established. Dissonance theory still lives.

versus social scientists, although the matter is sometimes presented in this over-simplified way. In his book, Schramm includes a section from a psychiatrist, Lawrence A. Freedman,[104] but Wertham has little sympathy with Freedman either. Yet Wertham, although undoubtedly one of the most outspoken, is not alone amongst psychiatrists in holding that television has detrimental effects. It is worth noting that of all those professionally concerned with the problem it is from the psychiatrists that we have the most unequivocal statements about the damage caused by television.

The statements from the psychiatrists may be unequivocal but they are rarely supported by the type of evidence which the social scientist would regard as acceptable and it is this question of acceptability that is one of our chief problems. Wertham writes of his "200 unselected cases" but his work does not merit an inclusion in either of the UNESCO bibliographies previously discussed. This is no doubt justified, at least from the point of view of the editors of the bibliography on the effects of television, who required that in order to merit inclusion an article or book should "reflect at least elementary standards of scientific research".

Whether or not this principle was consistently applied throughout the bibliography is not always clear. This question seems particularly relevant to some of the references to clinical work, general reports and conference papers. The concluding comment at the end of one abstract reads: "Unfortunately the research to which he alludes in the paper has apparently never been published."[105] It is not clear whether or not this piece of research passed the 'reflection' test.

In drawing attention to the questions of what is acceptable and what might be rejected and to the contending positions (there are, of course, further divisions both between sociologists and social psychologists, and within each of these disciplines), it is easy to exaggerate the differences and give an impression of total chaos and confusion. The impression can also be conveyed that there is no adequate base from which we may judge whether or not the results from a given piece of research should be taken seriously and considered in our general discussion.

It would be misleading if we painted too black a picture, for some criteria can be applied to the research and at least certain 'findings' can be safely excluded from our discussion. But the position is far from satisfactory, and whereas there may be a good measure of agreement on principles and criteria, there are many specific cases when agreement as to whether or not the principles have been applied, or the criteria observed, is not nearly so widespread. Small wonder that the layman is often bewildered by conflicting results and different evaluations of these results which social scientists frequently place before him.

It has been argued elsewhere* that, in addition to methodological advances, theoretical developments and conceptual refinements will be necessary before we can improve on this position. We shall have a little more to say about this in the discussion which follows.

One further general point can be made at this stage before we go on to discuss what is one of our central areas of concern, namely, television's portrayal of violence and aggression. We refer to the fact that when we have been discussing the effects of comics, films and television, we have stated on occasions that researchers have reported that the media did not have *a direct or causal effect* with regard to delinquent behaviour and that other factors would be operating in any delinquency situation. It is evident that here as so often in social science the word 'causal' is not used consistently, neither are the words 'direct', 'main' and 'major' which often accompany it.

One does not wish to create difficulties and it is quite likely that the general sense of the statements where such words as 'direct cause' are used is largely understood. However, in these cases there seems to be a possibility that some people will conclude that if television is not a main cause but *only a contributory one*, then it can be dismissed and need not be studied further. It will be remembered that we were faced with a similar situation when discussing the 'causes' of delinquency.

This paper is not the place for a full discussion on the nature of causal explanations in the social sciences, but we feel that we ought at least to draw attention to a possible source of confusion and misunderstanding. To put the matter simply: as far as delinquent behaviour is concerned ought we to expect to find that television could be anything but a contributory factor, a factor which necessarily works together with other contributory factors? This is mentioned here because of the tendency in some reports to put up the straw-man of direct cause (ill-defined though it may be), then knock it down (which as we have seen is not too difficult), and finally to play down the possible effects of television with a 'only a contributory factor' conclusion. It is with television as a contributory factor that we are primarily concerned.

We are entitled to hold sociologists, social psychologists and psychiatrists responsible for the quality of their research, but it would

* See, for example, the various publications of the Television Research Committee.

be unfair to hold them entirely responsible for the way committees, commissions, governments and the public at large interpret and use the results of their research.* We shall now go on to discuss the television–violence relationship, and in doing this we shall pay attention not only to the relevant research findings but also to some of the ways in which these findings have been interpreted and used.

Frederick Emery and David Martin[106] reported in 1957 on their enquiry into psychological changes induced by a 'Western' in a small number of Australian schoolboys between the ages of 10 and 13 years. Their results did not confirm what has sometimes been called the Feshbach hypothesis (this stemmed from Feshbach's work reported in 1955[107] that the experience of viewing television helps children to reduce their level of aggression). Their suggestion was that children developed a perceptual defence to protect them against the shock and anxiety experienced from violent films.

The ten years or so immediately following Feshbach's work produced quite a number of studies which dealt with the television or film portrayal of violence and aggression. The controversy as to whether watching violence on television was harmful, of little consequence or perhaps even beneficial for some became an important issue. Most of the studies which were reported were laboratory studies, yet extrapolations were frequently made to real life situations, and the 'findings' used, often without due caution, in the general debate. Some of the research centred on the imitation of aggressive models, whilst other projects dealt with the arousal of aggression or with the lowering of children's inhibitions with regard to it. The names most frequently associated with this work are those of Leonard Berkowitz and Albert Bandura. Statements from both of these psychologists are included in full, each of them meriting an appendix in the Interim Report[108] (Television and Juvenile Delinquency) of the Senate Sub-Committee to investigate juvenile delinquency in the United States which was made available in 1964. Similar evidence was put before the hearings of the National Commission on the Causes and Prevention of Violence (USA) in 1968-9.[109]

There can be little doubt that this experimental work has had considerable influence over the last ten years or so and, although it is not necessary to refer to all of the several relevant experiments (adequate summaries are available elsewhere[110]), we need to look a

* Some social scientists are more at fault in this respect than others. Some of those giving testimony before commissions seem to invite or encourage people to make generalizations from their work which, to say the least, are questionable from a social scientific standpoint.

little more closely both at the work itself and at the way it has been interpreted and otherwise used.

If as we suggested earlier social concern is reflected in the research effort then by comparing the two UNESCO bibliographies we can see that over the period in question there was a growing concern about the problem of television violence. Although delinquency research as such received if anything less detailed treatment in the 1964 bibliography than it did in the earlier one published in 1961, this was certainly not the case with the research on television violence. This had earned a separate section by 1964.

Not surprisingly, it was mainly the new wave of experimental work that had attracted the attention of the Editor, and although the limitations of this kind of work at least as far as generalizations to real life situations are concerned are fully recognized, the work seems to be highly valued. It is suggested that it is the experimental method that is most likely to enable mass communication research to move from the plateau on which many people thought that it had rested for far too long.

The general conclusions reached in the UNESCO publication are that there can be little doubt that violent programmes on television do not serve to reduce aggression vicariously but, if anything, increase it and encourage its later expression. It is maintained that there are particular dangers for those children who come to the television with high levels of aggression, for the portrayal of violence may feed that aggression as well as provide hints which could facilitate its anti-social expression. Attention is also drawn to the dangers which might be associated with regular identification with aggressive heroes and, as so often in the past with the other media, to the high vulnerability and susceptibility of certain groups of maladjusted and friendless children.

A study by Riley and Riley,[111] in addition to pointing out how the child's relationships with his friends might help to determine how he reacts to violent programmes, also introduces the fantasy–reality dimension. The suggestion is that a child with unsatisfactory relationships is more likely than others to confuse the border of fantasy and reality and to compensate for the inadequate relationships by making use of violence learned from television.

There can be no doubt that the child's ability to make the distinction between what is real and what is fantasy can be crucial and we have already seen that television and the other media can be used in different ways by different people. There is a danger however that an over-simplified view of the research results which support conclusions such as those reported by the Rileys can lead to crude 'theories' of substitution and compensation. Exclusive concentration on the child–television relationship can be misleading, for it is a relationship which needs to be set and examined in a wider context. Relationships are more complex than any one-to-one equation would suggest. Whether the disturbed or friendless child automatically turns to television for his compensation seems doubtful. Some may, others may not. The actual direction adopted will obviously depend on a much wider range of experiences, interests, relationships and opportunities. In the widest possible sense it depends on what is available.

One further word of caution is necessary about the interpretation of research results which mark off a particular individual or group in terms of heavy viewing, specific programme preferences, identification with television heroes, and so on. As we shall see later, this sort of information *by itself* does not necessarily tell us anything about the direction of the relationship or about the personal or social consequences of that particular pattern of viewing preference or identification. Heavy viewing and delinquency could both be symptoms of the same wider condition. They may feed each other but, on the other hand, they could be seen as alternative or competing outlets.

Returning to the question of violence and to the UNESCO bibliography it is important to note that, despite all the qualifications and limitations, it was still considered that on the whole, the weight of the evidence was behind Berkowitz's conclusion that "the heavy dosage of violence in the mass media", although not a major determinant of crime or delinquency, "heightens the probability that someone in the audience will behave aggressively in a later situation".[112]

As we have seen, great caution is required in making any generalizations to real life situations from the type of experimental design used by Berkowitz, Bandura and others. Yet results from this type of work are frequently used to support the type of statement about the effects of television like the one just quoted. Both Berkowitz and Bandura have been criticized for lack of customary caution in this respect and there does appear to be some justification for this criticism.

What is probably more important is that their work has also been subjected to more serious criticism. A report produced by Ruth Hartley,[113] under the auspices of the Columbia Broadcasting System's Office of Social Research, draws attention to weaknesses in methodology, design, analysis and conceptualization and, on occasion, questions the statistical significance of the findings as well as some of the general inferences.

These criticisms and their substantiation do not lend themselves to brief summary, and cannot be pursued further here. However, notwithstanding the interests of the sponsors of the report, its general defensive orientation, its own inadequacies, and the fact that Berkowitz in particular has developed his work since the report was published, the criticisms have to be taken seriously.

Leo Bogart, well known for his research in mass communications, would not apparently accept all these criticisms, but he shares with us a concern about the general application of such results. Bogart also shares our feeling that in the discussion on media violence far too much attention has been given to the matter of direct imitation. This is not a major area of concern. In his evidence before the National Commission on the Causes and Prevention of Violence, Bogart had this to say:

> "The experimental evidence regarding the effects of media violence on children's subsequent behavior is to me entirely convincing so far as it goes. However, a great deal of experience in the measurement of media effects indicates that it is far easier to trace them in the pure experimental conditions of the laboratory than in the natural conditions of the field, where attention to the message is not heightened and focused as it can be in the laboratory, and where the pressure of competitive messages and activities comes into play. In spite of the fact that communications effects are more attenuated when studied under natural field conditions, I know of no instance where laboratory effects are *reversed* in the field.
>
> "The inference to be drawn from this is that if exposure to violence increases the subsequent display of aggressive behavior among experimental subjects it tends to move people in the same direction under normal exposure. But there is no way that I know of to infer just what the level or intensity of that effect will be, relative to the laboratory effect. We know that it is apt to be different for different kinds of people, so one variable to be considered is the composition of the sample of subjects, compared with that of the true population. Just as children may react with greater intensity than adults to the same media content, so slum children may react very differently than middle-class children.
>
> "The effect of a communication can never be gauged except with reference to the susceptibilities of the audience. Advertisers expose their message to vast numbers of people, knowing well that only

a small handful are potential buyers, interested in the product offered. The individual who is looking for kicks of any kind can find them in the media by going out of his way. A small minority of disturbed or angry individuals may find stimulation in media depictions of violence, but if the media did not furnish them with models of violent behavior, it seems farfetched to suggest that no other models would be available."[114]

In passing, it is interesting to note the reaction of the television industry to any research which purports to show that watching violence, or any other television experience, may have undesirable consequences for the viewer or society at large. One sometimes gets the impression that in the United States the media men have 'experts' on hand whose main task is to counter the statements and look for weaknesses in the work of any other 'expert' whose work may be interpreted as an attack on television. Of course, if there are weaknesses in research it is important and necessary that they should be identified and exposed. However, we must remind readers about what was said earlier on the nature and general state of social scientific enquiry in this field. It is not that we wish to stand in the way of a rigorous examination of research methods and results, although we would be happier if we were convinced that the rigorous examination was consistently and uniformly carried out, and if we felt that the interests of social science or society at large rather than those of the television industry were at the heart of the matter.

The television industry in the United States has not exactly been to the forefront in the disinterested pursuit of knowledge about the operation of the medium and its effects. It simply will not do to argue, as one senior television man is reported to have done before the National Commission on the Causes and Prevention of Violence, that research in this field had not been well supported by the television industry because a foolproof method was not available. The best way, in fact perhaps the only way, to improve methods is to carry out research.

Even in England the reaction of the broadcasting men to the television–violence controversy is not without interest. The Chairman of the BBC, Lord Hill, is normally associated with what has been termed a commonsense approach to broadcasting and is not particularly well known for his interest in research. Yet on one of the few recorded occasions when Lord Hill[115] has mentioned a specific research project in public it was to refer to some incomplete, and at that time unpublished, studies by Professor Seymour Feshbach which, so it was claimed, gave some support to the cathartic argument and threw doubt on the idea that violence created violence.

Returning to our main theme, perhaps we should attempt to keep the record straight by making it clear that Feshbach is not the only person to offer evidence in support of catharsis. In addition to other studies from the United States, support is also forthcoming from Italy and France.

Work from Australia[116] where distinctions were made, both with regard to types of violence portrayed and types of response produced, indicate that a "more subtle form of catharsis appears to be a possibility". The child and the adolescent project their personal, even subconscious, problems onto the film and the viewer is provided with the means of expressing and translating his conflicts. 'Catharsis' here results from the conjunction of several mechanisms, some dynamic or expressive, others inhibiting. It also seems possible that the child can develop a perceptual defence to protect himself from the shock and anxiety experienced from violent films.

This work also suggests that children who view crime dramas might acquire an insensitivity and stereotyped reactions to violent events in real life. It confirms that wide generalizations about children in general and violence in general are not likely to be valid and it lends some support to those who stress the importance of taking into account such factors as context, plot and genre in the study of effects. André Glucksman,[117] writing in France, and surveying the whole field, accepts the possibility of catharsis for some people in some situations, as he also accepts the possibility that for other people media violence could induce real violence by identification or the dissemination of knowledge. He concludes, as so many others have done before him, that there is no 'direct effect' on behaviour. More significantly, he draws attention to the fact stressed earlier by one of the present authors,[118] that it can be extremely misleading to attempt to deal with what is an extremely complex behavioural phenomenon by the use of a single concept, 'violence'. The Television Research Committee has also suggested that we need to ask both with regard to stimulus and response: What is violence? What is aggression? We also need to know how different forms of violence are perceived.

After carrying out what is probably the most comprehensive survey to date on the media–violence question Franco Ferracuti and Renato Lazzari[119] state that the cathartic effect has not yet been confirmed by the results from any acceptable piece of research. The available evidence, so they say, offers more support for the hypothesis of incidental learning. They feel this is particularly true for those children who are more vulnerable than others. They recognize that little can be said with certainty about identification, and even less about the long-term effects of television on attitudes and values.

Addressing themselves directly to the question of public concern the authors of this survey state that, as far as 'normal' children are concerned, the widespread fears are not justified—at least not on the basis of the available evidence. Ferracuti and Lazzari accept that violence in television and the other media can be dangerous for the abnormal and the maladjusted child* and, in addition to asking for a greater research effort to help solve the problem, they also call for a policy of self-censorship and careful selection in the presentation of violent material.

The most up-to-date report on the effects of television violence comes from the United States where, in September 1969,[121] the National Commission on the Causes and Prevention of Violence issued a Statement and made policy recommendations. In discussing this Statement, it should be remembered that it is not a statement by social scientists, but by a Commission which, after taking evidence and guidance from social scientists, media men and others, drew up its own conclusions and recommendations. We may also assume, with one or two possible exceptions,† that the Commission had available to it the same sort of research evidence that had been available to Ferracuti, Glucksman, and the Television Research Committee in Great Britain when they had been reviewing the field and drawing their conclusions. One final point about the Commission's Statement, and that is that it obviously deals with the American situation where there is more violence both on and off the screen than there is in this or in most other 'developed' countries. From our point of view, the main interest in the Statement is not in its references to the quantity and quality of violence on the American screen, nor is it in the recommendations about changes in specific types of programme or general programme policy; it is in the processes by which it is alleged that television has its influence and effects.

* One of the present authors suggested several years ago that an important research task in this area would be to identify the nature and extent of these vulnerable groups in our society.[120]

† Work available to the Commission but not generally available when the above was written included content analysis on the dimensions of violence in television drama, and a survey of the real world of violence as experienced by Americans. Both studies were carried out on behalf of the National Commission's Mass Media Task Force and, together with other evidence, are reported in the Task Force's most comprehensive report to the Commission[122] published after the above was written. An edited version of the Statement appears in the Commission's Final Report.[123]

It would certainly appear that the members of the Commission were not impressed by the evidence produced by those who supported the cathartic position, for it is argued in the Statement that there is little support for catharsis or 'draining off'. In fact, it is clearly stated that televised violence stimulates aggressive behaviour and that aggressive behaviour can be learned from television and retained over time. The emphasis seems to be on learning, particularly on incidental or observational learning which, so it is maintained, depends on the degree to which the child can identify with television characters as well as to the extent to which he can perceive utility or anticipate gratifications from his viewing experience. The younger child between the age of three and eight years is seen as being particularly vulnerable. The inexperience and immaturity of the child at this age, the fascination new material has for him and his general inability to make clear distinctions between fantasy and reality are all considered to be important in this connection.

Learning is also mentioned with regard to adolescents who "consciously rely on mass-media models in learning to play real-life roles".[124] Moral and social values can also be learned and it is said that children learn from television because it provides them with the "most accessible back door to the grown-up world". It is feared that the image of the adult world which most children get from television is by and large an unwholesome one. This last point of course depends very much on what content is provided by television and this obviously varies from country to country.

These are relatively straightforward and unequivocal assertions about the undesirable influence of television. They tend to be more sweeping, more general and less qualified than the assessments of the Television Research Committee. The contribution of other non-television factors in the learning and general socialization process and the complexity of the social settings are recognized and accepted and, even in this document, we still find the customary qualification "we do not suggest that television is a principal cause of violence in society".[125] Nevertheless, it is firmly stated that television is a *contributory factor* (making a considerable contribution) and that "it is a matter for grave concern that at a time when the values and influence of traditional institutions such as family, church and school are in question, television is emphasizing violent anti-social styles of life".[126]

The conclusion is "that a constant diet of violent behaviour on television has an adverse effect on human character and attitudes—encourages violent forms of behaviour and fosters moral and social values about violence in daily life which are unacceptable in a civilized society".[127]

The National Commission Statement deals only with television entertainment programmes. It contains no specific recommendations about non-fictional material. We recognize that the work on violence in television entertainment is but one aspect of a much wider study of violence in American society. There have been other statements from the Commission, and its Final Report contains sections and recommendations on Group Violence, Violent Crime, Assassination, Firearms, Protest, Campus Disorder, and so on. The news media, however, do not figure prominently in any of these sections, or in the Report as a whole. There may be good reasons for this which are not immediately obvious to the outsider, and we can readily accept that it is easier to make recommendations about television entertainment than about news coverage—the conflict with free expression is not so acute in the former case.

We certainly cannot ignore violence in entertainment programmes, but to confine recommendations (changes in programme schedules etc.) to this type of programme, as the American Commission has done, could be misleading. The imbalance could lead to an inadequate definition of the problem, the recommended action could be seen as making a real contribution to the reduction of violence in society, whilst other more significant media forces in non-fictional presentation might be ignored. If the problem is wrongly diagnosed, the chances of a successful solution are small.

The Commission shows that it is aware of the complexity of the situation, that television entertainment cannot be a root cause, and that various explanations—ranging from innate aggression to the widespread frustrations of urban living—are available to account for the existence of violence in society. The Report is wide in its general coverage, but relatively narrow when it comes to the mass media. It is also much narrower in scope than the Report which was submitted to it by the Media Task Force. This Report does cover news, and makes specific recommendations about research policy and institutional development. We do not know what factors governed the Commissioners' selections from and interpretations of the Task Force Report, but it would appear that many of its recommendations were not accepted. It is just possible that overstatement of the case did not help.

When the work of the American Commission is compared with the *Second Progress Report and Recommendations of the Television Research Committee*[128] the differences in the terms of reference, the nature of the public dialogue, and above all the fundamental historical and social differences between the two countries concerned, must be given due weight. The Television Research Committee, with wider and less specific terms of reference (at least with regard to violence) than its American counterpart, draws attention, amongst other things, to television's portrayal of realistic, more generally acceptable or 'legitimate' forms of violence such as war, punishment and self-defence. The possibility that even straight news reporting may have detrimental effects is also recognized.* It is implicit in some passages in the Report that if day after day young people get the impression that it is all right for society to use violence to solve its problems, then they might possibly feel that it is all right for them to use the same methods to solve their problems.

The Report of the Television Research Committee is similar in many ways to, and shares a great deal of common ground with, the American Commission. The Americans seem, however, to have been more impressed than the British Committee with both the results and the potential of the experimental approach as described earlier in these pages. The American Commission appears to have more faith in the laboratory work on imitation, arousal of aggression, and the lowering of inhibitions. More central to the thinking of the British Committee are the long-term effects in the shape of desensitization, over-familiarization, changing patterns of acceptance, increasing tolerance of violence, increased readiness to think of violence as a solution to problems, and endorsement of belief systems in which force plays a central part.† The Committee also draws attention to the possibility of violence on television leading to anxiety and other forms of distress and suggests that it is a mistake to confine enquiries to the violence-producing-violence question.

Perhaps the main differences between the two bodies are accounted for by the fact that the Television Research Committee's terms of reference were not confined to entertainment programmes. However, whatever the differences, and these are only incidental to our main concern, the fact is that in our discussion of the relationship between television and violent behaviour we cannot ignore the way television and the other media deal with the nature and occurrence of violence in real-life situations. What pictures of violence are presented to us?

* It is also accepted in the Report that the portrayal of real violence could have beneficial effects, for example, by deglamourizing war and possibly by increasing the social visibility of certain social problems, e.g. civil rights in the USA.

† These effects were not ignored by the American Commission.

What stereotypes, definitions and classifications of violence are put before us? What forms are approved—what disapproved? What kinds of violence are made an important issue or defined as a social problem, and when does this occur? When is violence offered as a solution? Does television over-state or amplify the problem? Are some groups influenced, perhaps given an identity, by the way in which the media over-expose them? Do some groups find excuses in television content for their behaviour? These are the vital questions. We do not have to deny the importance of identification, imitation, or other more conventional approaches, as described in the American Commission's Statement and elsewhere, to hold that answers to the questions just posed are the ones we must try to obtain. Such answers would throw more light on the nature of our social problem than those based on imitation and other over-simplified models that have so often been used in the past.

In attempting to find out if there is any relationship, or even the possibility of a relationship, between television and delinquency, we looked first at delinquency research and then at one particular aspect of mass communication research, namely, the research which addressed itself directly to the relationship. We said earlier that there were bound to be overlaps between the two areas and that there would also be some overlap between the type of mass communication research just covered and mass communication research in general.* It could be argued that in this book there is no need to make this last-mentioned distinction, that most of our distinctions have a large element of arbitrariness about them, and that this one adds nothing to our general exposition. Such an argument is not without substance but we do feel that as mass communication research is central to the whole subject, it would be a mistake, and could mislead our readers, if we gave the impression that the research referred to so far was representative of mass communication research as a whole. We do not propose to deal with this wider area in any but the briefest and most general way, for adequate summaries are already available.† However, we hope that in the short discussion that follows we shall not only correct any imbalance that may have been created, but we might also be able to point the way to some of the weaknesses and shortcomings in the research, and indicate the additional questions that we must ask if we are to improve our understanding of how the media influence people. It is appreciated that in doing this there will

* Even when using these terms, the discussion is confined to effects research which is itself only one part of mass communication research proper.

† See, for example, the publications of the Television Research Committee.

be some repetition of some of the general points made in previous pages.

We shall proceed quite simply by stating (without necessarily endorsing) what appears to be the generally held position on the effects of mass communication, and then go on to add some qualifications.

Television cannot be studied usefully in isolation. It seems to be generally accepted that, in certain circumstances, television and the other media can influence values, attitudes and behaviour—but that, in most cases, television is unlikely to be operating in a vacuum, and that its influence will be exerted only in conjunction with a variety of other factors. The television experience is studied by some social scientists as one aspect of an ongoing process of social interaction, and it has been seen by others as a single strand in a complex network of social relationships. Approaches have also been adopted which entail studying or tracing the message as it enters and spreads through an already existing communication network in which people are in different places and play different roles with regard to the message.

Individual differences on a variety of scores, and the accompanying differential susceptibilities and vulnerabilities, are recognized and generally taken into account. The individual is not seen as a *tabula rasa* in front of the screen but as a person with past experiences, present affiliations, and future expectations—all of which play a part in determining the influence of any given television experience. This form of thinking has led social scientists to ask questions which are framed more in terms of the different uses which different people make of television than in terms of what television does to people. The predominant model is one of exchange and interaction between medium and audience, rather than one which assumes exploitation by an all-powerful medium.

It seems to be generally accepted that as far as the influence of television is concerned, it is more effective in conveying knowledge than in changing attitudes. This is not surprising. Neither is the fact that—once they are well formed—attitudes, particularly those in which people are highly involved, are difficult to change. Little can be said with certainty about the long-term or accumulative effects of television on attitudes and values but, on the whole, television is thought of as reinforcing what is already there, rather than creating new ideas or changing old ones.

The effects of television can be experienced at several levels, from leisure-time behaviour patterns to gains in knowledge and changes in attitudes and values. The importance of dramatic presentations, repetition of given themes or values over a period of time, an uncritical attachment to the medium, and an absence of standards and references that could be brought to bear on the television experience, are all thought to contribute to impact and effect.

The above are generalizations, from the work carried out over a period of twenty or more years. They obviously depend on the type and quality of the research from which they stem. As they are essentially answers to questions that have been asked, it is not only important to look at these questions and how they were asked, but it is also necessary to take account of the questions that have not been asked. Our research results can never be better than the questions we ask in the first place. We should always bear in mind the possibility that if a relationship has not been established it could be as much a function of the research question and design as a reflection of the real situation.

Some of the major criticisms which have been made about mass communication research have been implicit in much of what has been written earlier in this book. We can only draw these together here and do little more than list them. They must be taken into account, however, when we are evaluating what has been 'established' in this field of research.

Lack of theory, and the tendency to do rather than to think, have been two of the main criticisms. It would appear that mass communication research is better known for certain aspects of its methodology than for its refined conceptualization. In discussing violence, we have already seen some of the problems which can accompany a lack of conceptual precision and refinement. In reviewing the general field, it is also possible to gain the impression that mass communication research consists of many discrete piles of information, unintegrated, and often apparently unlinked in any way. The predominance of administrative, service and commercial research has often been blamed for this state of affairs. Culpability need not concern us here, but it does not take long to realize that many of the research projects do not appear to have been informed from a theoretical position, or to have stemmed or benefited from a developed corpus of relevant knowledge.

Another criticism frequently heard is that mass communication research, in assessing effects and influence, has relied too much on the concept of attitude. Moreover, it can be argued that the model of attitude change, which it has been customary to use, has not been capable of picking up many of the changes that might possibly come from television. This is likely to be particularly true of long-term effects, where quite substantial changes might come about by a gradual process of accumulation over time; yet at any given stage in this process, a 'before and after' study on one particular programme might show nothing at all.

Many studies using a variety of methods have attempted to measure attitude change following exposure to a particular programme or series of programmes. Often, although not always, this type of research has dealt with specific messages, information campaigns, and so on. Because of this, there has been a tendency to confuse effects with effectiveness. An absence of change in the expected or intended direction has been interpreted as an absence of influence. As we saw in our previous discussions, there is much more to the influence of television than can possible be assessed through direct changes in attitude and opinion, as these have normally been measured. The concentration on attitude change as a primary criterion of influence has meant that other important areas have been neglected. Television can help set standards, confer status, define norms, provide stereotypes, convey a sense of what is approved and what is not approved, structure our fields of discourse and frameworks of expectation, and generally provide us with our pictures of the world. Influence must not be equated with attitude change and, in general, the tendency to do this in the past has probably led to both an underestimation of the extent and a misunderstanding about the nature and direction of the influence of television.

In assessing the generalizations about the effects of mass communication, it needs to be appreciated that very little work has been carried out on the role of television in the early stages of the child's development. It may be true that television has relatively little influence when attitudes and values are well formed, but what about when they are not well formed, but are actually in the process of formation? The role of television in the lives of very young children particularly with regard to the confusion between fantasy and reality should be a priority research area.

Obviously the family is one of the primary agents in development, and the closeness and intensity of family relationships are usually regarded as vital aspects of the process. It is sometimes suggested that the impersonal nature of television precludes the development of such relationships between child and television, and that therefore the medium should never be regarded as a socializing agent on a par with the family or school.

We can be too restricted, however, in our approach to child

development. The child changes, and the society in which he grows up also changes. Socialization must always be studied in terms of a developing child and a changing society. It takes place in interaction with many people and in several settings, and the 'significant others' in the process are certainly not confined to any one agency. The family gives way to the school and to the peer group; in varying degrees teachers and friends take over from mothers and fathers as the main reference points and as the people who matter. Is there a place for television in this process?

We have partly answered this question in our discussions on delinquency, when we tried to see if television had been or could possibly be a factor contributing to delinquent behaviour. It was suggested then that if television did play a contributory part in the development of delinquency, it would probably be indirectly, perhaps through inter-personal relationships and social settings. The same thing applies when we are discussing socialization, which is not surprising, for we are of course talking about the same thing.

For example, we might learn something from a television personality whom we regard as a model, and what we learn could become part of our store of knowledge which may or may not be used. On a suitable occasion it might be tried out and tested in our everyday activities. If the trial is successful, i.e. if it works and we are suitably rewarded, gaining some sort of satisfaction, then this could act as a reinforcement, and perhaps we could then talk about the development of a pattern of influence.

The possibility of television having an indirect influence can be illustrated further by reference to the situation where the medium presents a certain attitudinal or behavioural pattern as being the appropriate style, or the 'done thing' in a specific social group. The approved and disapproved behaviours, the requirements for membership, the rewards and the sanctions, may be spelled out in a variety of ways. For some members of the audience this group could be a highly valued reference group, and we may assume that such people would wish to conform to the group norms, to be accepted, to have status, to match up to the requirements, and to make the grade. It is in this way, showing what has to be done in order to achieve a valued goal, working together with other forces and through other agencies, that television might have its influence.

Earlier, when approaches to the study of delinquency were examined, we suggested several ways in which television might be used in connection with delinquent behaviour. Two of these ways, namely, the learning of 'techniques of neutralization' and the use of television in the 'differential association' learning process, spring immediately to mind—and both are referred to again with regard to future research possibilities at the end of the research report in the second part of this book.

These by no means exhaust the list of possibilities. Once we free ourselves from thinking in terms of the conventional models of attitude change and direct effect, we can see that it is *possible* for television to do many things. For example, it may define problems, outline situations, present models, limit choice, extend experience, provide knowledge or stereotypes, reinforce one's self-image, confirm expectations, confer status, indicate goals, stimulate aspirations, and foster popular culture and leisure goals.

It is not difficult to see how these possibilities link up with those previously listed when we discussed delinquency. Perhaps one additional point can be made, and that is that where much of the experimental work has looked at the reactions of artificially frustrated individuals to televised or filmed aggression, very little attention has been given to television as a source of frustration. There is no one clear way or approach in research that is better than all others. But in studying violence in society we feel that it might be more fruitful to study television as one amongst several factors which may act as a source of discontent, over-stimulating aspirations, and helping to produce anxiety, tension, frustration and possibly violence, than to study for example direct behavioural imitation of televised models.

This is, of course, much easier said than done—and we must repeat that much of what we have just said is essentially speculative. We think there are good grounds for our stance, but we readily recognize that the research results reported in the second part of this book do very little, as results, to advance our position. However, this research marks a beginning—no matter how small—and it is the leads that have come from it, the more focused approach that we can now adopt, and the hypotheses that we feel can now be tested—that represent the real gains. We at least know now, from what is the first study in this country of delinquents' television behaviour, that there is little to be gained in studying delinquency or television in isolation. Television behaviour should be seen as but one aspect of a wider media behaviour which, in turn, is but one aspect of a wider leisure pattern — and even this wider pattern can be adequately understood only within the still wider social setting of our society.

Part Two

THE STUDY

I

INTRODUCTION

Research paradigms

The research design employed in the present study is essentially correlational: that is, analysis of the data seeks to show whether or not there is any correlation between delinquency and certain aspects of mass media behaviour. In the present case, delinquency was 'measured' on a nominal scale, in the sense that respondents had either been legally categorized as delinquent, or not. In addition, it should be pointed out that adolescents who had committed more serious offences were unlikely to be included in the sample of delinquents studied here, since such offenders are liable to be sent to remedial institutions rather than placed on probation. Mass media behaviour was measured in a number of different ways. The questions asked and the manner in which responses were scored are discussed at appropriate points in the following sections.

As a means of studying the effects of the mass media in causing or controlling delinquency, the correlation study is, of course, only of limited utility. Even if we find that there is a strong positive correlation between delinquency and certain types of media exposure or usage, this still tells us nothing about the causal sequences which underlie our data: without further evidence three major possibilities then achieve equal likelihood. In the first place, delinquency may have caused the observed pattern of media behaviour; alternatively, media exposure may have caused the delinquency; or, thirdly, both the delinquency and the particular patterns of media behaviour may have been the result of some other, possibly unlabelled, and perhaps unknown factor. Apart from these three straightforward and indeed somewhat naïve formulations, we can of course imagine other sequences of events which would have produced the positive correlation we have observed. For example, those prone to delinquency may already have manifested unique media behaviour patterns prior to their committing a first delinquent act, with an intensification of such patterns *after* committing such an act and being labelled as delinquent by society. But however far we speculate, purely correlational data will not allow us to choose amongst these alternatives, though related theory and related findings may make one seem more likely than the others. On the other hand, if we find that there is *no* correlation at all between delinquency and media behaviour, then we shall be fairly safe in rejecting all of these possible causal hypotheses and concluding that delinquency and media behaviour are not related in any meaningful way.

This brief discussion may make it seem as though purely correlational studies are hardly worth doing at all, since they will almost inevitably leave major questions unanswered at the end of the day. But of course correlational studies often have to be resorted to because it is both practically and ethically impossible to conduct experimental studies with human subjects in real-life situations. For example, we might ideally like to take a sample of young children and allocate them randomly to a number of experimental groups. We would then control very precisely the media materials available to the different groups over a period of several years, and make measurements at the end of the period of the actual or potential rates of delinquency characteristic of the different experimental conditions. But even if this were ethically possible (which it clearly is not), the practical difficulties would prove virtually insurmountable: the mass media are ubiquitous, and there would be little real chance of effectively controlling the exposure patterns of the various groups.

The researcher is thus forced back to correlational or survey studies of one sort or another, unless he is to content himself with laboratory experiments. However, some of the many drawbacks to researches of this latter kind have been mentioned already in the first part of this monograph.

By being fairly liberal in what is included, quite a lengthy list of correlational or survey studies of the mass media and delinquency can be assembled. Many of these studies date from the 1930s and 1940s. The majority are described and summarized by Pfuhl, and some discussion of work of this nature has already been offered.[1] In the present context, however, we must draw attention to several specific factors which limit the relevance and utility of these earlier pieces of research. In the first place, few of the studies deal with television. Mass communication research suggests that television is in several ways functionally equivalent to older media. Even so, television's ready availability and distinctive content, and in particular the amount of leisure time commonly devoted to viewing, all point to the need for studies focusing specifically on this medium. Second, much of the available research was conducted in the United States, and this can clearly mean that the relevant results are not directly applicable to the British scene. Quite apart from the obvious social and cultural differences, it is clearly unreasonable to equate American media output with media output in this country.

A third and yet more serious shortcoming of much of the available research, and one that has something to do with the time period when it was conducted, is the fact that few of the studies are based on sociological or social-psychological theories about crime, delinquency or deviant behaviour. This is not surprising, since the relevant sociological theories, particularly those now usually known as subcultural theories, date from quite recently: they were simply unavailable to investigators working in the 1930s.[2] But this does mean that much of the available research is outdated at the theoretical level. The present study was not based in a very close way on sociological theories of delinquency. However, reference is made to several of these theories in an attempt to explain the findings. Despite these limitations, there are three American correlational studies which parallel the present research particularly closely, in the sense that each compares the media behaviour of a delinquent sample with that of a matched non-delinquent group. Only one of these involved the study of television, however. But an analysis of some of the problems raised in the interpretation of this prior research may be helpful later on in considering the strengths and weaknesses of the present investigation. We suggest that some of the crucial aspects of research of this sort are: the way delinquency is defined; the recruitment of the delinquent sample; the extent and nature of the matching between delinquent and non-delinquent samples; the range of mass media studied; and the depth at which media behaviour is probed. Questions about several of these aspects are raised by both our own research, and by the most elaborate of the American studies. We turn first to look at this piece of work in some detail.

Prior research

The most extensive piece of survey research designed to elucidate the relationships between the mass media and juvenile delinquency was conducted in three urban areas located in the state of Washington in the United States. The investigation of the media correlates of delinquency formed part of a much larger survey of the social structural and other correlates of delinquency in that area. Pfuhl's dissertation reports the relevant findings.[3]

The sample was drawn systematically from a high school population comprising all those in the 9th to 12th grades, and finally consisted of 792 respondents, divided approximately equally between the sexes. The data were collected by means of self-completion questionnaires administered during school time. A large number of questions dealt with mass media behaviour, and these covered reading, cinema going and television watching.

Unlike the present study which relied on a legal definition of delinquency, Pfuhl made use of the Nye and Short self-report method to categorize his respondents as more or less delinquent.[4] Pfuhl presents a neat summary of the advantages of the self-report method over that relying on the categorization of juveniles as delinquent by the legal system. Even so, an examination of the items comprising the measuring instrument leave one somewhat unconvinced that what was being measured was delinquency as usually understood. It is true that the test attains very acceptable levels of reliability and internal consistency and successfully distinguishes between fairly extreme criterion groups; but some doubts remain. For example, two of the seven items finally employed for scoring purposes ("Skipped school without a legitimate excuse", and "Defied your parent's authority (to their face)") do not in themselves involve behaviour which would be subject to legal penalties if detected. In addition, one comes back to wondering whether the important distinction is not between those who go far enough in their activities to come to the notice of the police or other adult authorities, and those who do not. However, the use of the self-reporting technique allowed Pfuhl to sort his respondents into those with high, medium and low delinquency scores, rather than merely into two classes.

But quite apart from the question of how the data on delinquency proneness was *collected*, a doubt needs to be raised about the chosen approach to data *analysis*. Pfuhl divided his respondents into three groups—those with high, intermediate and low scores on an amended version of the Nye–Short test. Taking the tabulations for boys as an example, it is clear that the ranked delinquency scores were trichotomized in such a way that approximately one third of the respondents fell into each of these three classes. For unexplained reasons, the totals vary slightly from table to table, but taking three tables presenting details of the three statistically significant relationships between delinquency and media patterns, the maximum numbers of respondents are: high on delinquency, 117; intermediate on delinquency, 124; and low on delinquency, 145. Now, happy as this arrangement is for the calculation of statistical tests, one wonders whether it has much social meaning. If one treats approximately one third of one's population as delinquent (presumably those high on delinquency are generally meant to be equated with 'delinquents' in the usual sense), then is this very meaningful? If juvenile delinquency is considered as a form of social deviance, then one would have to presume a very extreme degree of social disorganization if one third of the population in fact showed such deviance. To put the matter in other words: would not Pfuhl's tabulations have been more realistic if he had taken a much smaller and more extreme group as his delinquents, and compared them with the rest of the sample? Although the trichotomization of the sample into three equal-sized groups is statistically convenient, and provides some guarantee against treating non-linear correlations as though they are linear, it perhaps bears too little relation to a social situation in which everybody involved thinks in terms of a small group of delinquents and a large group of non-delinquents, and acts accordingly. These considerations certainly need to be borne in mind in the assessment of Pfuhl's findings.

It is indeed a pity that the available data were not analyzed in a more flexible fashion, with various cutting points on the delinquency continuum being tried out, and in general a more exploratory and open-minded attitude adopted. Perhaps it would have been as well to recognize too that a random sample is likely to furnish only a small number of adolescents who have actually been 'caught', and that if one adopts the attitude that adolescents *in general* are liable to delinquency (as seems characteristic of Nye and Short), then the sheer commonness of the behaviour makes it unlikely that it will be strongly correlated with other factors. The behavioural sequence we label 'brushing one's teeth' will show certain class correlations, but rather weak ones: we may expect equally weak (and perhaps equally uninteresting) correlations if we adopt a 'weak' definition of delinquency.[5]

Pfuhl's actual research hypotheses are in fact quite complex, involving as variables the degree of parental rejection and the incidence of psychosomatic complaints, as well as delinquency proneness and mass media exposure. However, no attempt will be made here to present the full findings; attention will be focused on evidence concerning the major relationship under investigation.

In the reported tabulations, boys and girls are always considered separately. Without controlling for any other variables, and categorizing respondents into the three groups according to their high, intermediate or low delinquent involvement, Pfuhl finds three statistically significant relationships between delinquency and mass media behaviour for boys. Those higher on delinquency attended the cinema more frequently, saw more crime films, and were more interested in crime and mystery novels than those low on delinquency. Similar tests were made for girls, and here the only significant differences were in terms of frequency of cinema attendance and exposure to crime comics, with the more delinquent girls having a higher exposure to each medium. Since preliminary analysis of the data had shown that frequency of church attendance was significantly correlated with both delinquency and media behaviour, Pfuhl proceeded to re-examine these five positive correlations with church attendance held constant. When this was done, the strength of the previously discovered relationship between boys' interest in crime and mystery novels and their self-reported delinquent status fell below that required for statistical significance. The other four relationships, however, were not materially affected.

Given these findings, it is perhaps somewhat surprising that Pfuhl summarizes this aspect of the research by claiming that the primary hypothesis — that there should be a positive correlation between delinquency and media usage—has to be rejected. It is true that, in all, fourteen indices of mass media behaviour were used, and that the *majority* of these indices revealed no relationship with delinquent status for either boys or girls. But it is worth noting that only six of the indices had specifically to do with media content involving crime. Thus, taking the figures for boys and girls together, of the twelve most crucial tests (six crime-content indices × two sexes) four, or one third, were found to be significant when the relationship was examined holding religion constant. This is, of course, well above the figure that would be expected given the selected significance level of five per cent. One has therefore to conclude that Pfuhl is somewhat harsh in the criteria he implicitly accepts as relevant to the testing of his major hypothesis, and that in his somewhat mechanical handling of the results he fails to go very far into an interpretation of his findings.

Perhaps the fault in fact goes back to the original choice of those aspects of mass media behaviour to be investigated. While the sub-hypotheses involving parental rejection and psychosomatic complaints are quite sophisticated several of the indicators of media exposure and usage are relatively crude. Thus purely on commonsense grounds we might hypothesize that the delinquent and the near-delinquent would be particularly interested in crime films; yet we would hardly be so sure that there ought to be a comparable relationship between delinquency and 'time spent listening to radio', another of the quantitative indices based on the questionnaire material. Indeed, one can go further and argue that crude measures of exposure (such as amounts of time spent listening to radio or watching television) are least likely of all to correlate significantly with some other social variable, since consumption of these electronic media in the home is known to be so all-pervasive. (The same argument probably does not apply to the cinema, and certainly the findings here would lead one to reject the argument, if advanced.) It is hardly surprising, in other words, that Pfuhl found that only some of his media indicators differentiated between groups of respondents scoring differentially on a test of delinquent behaviour. What is interesting, and what tends to get lost in the summary of the research, is that several of the key indicators *do* in fact differentiate between delinquents and non-delinquents.

Further, Pfuhl has at his disposal very little data on how his adolescents actually used the various media, due in part no doubt to the constraints imposed by the use of self-completion questionnaires and a large sample. He has questions which provide measures of amount of exposure, both to different media and to particular (though broad) categories of media content, and also some which indicate preferences for crime content at the cinema, on television and radio, and in books and comics. But such questions tell us little about how different classes of respondents actually use the content to which they expose themselves: that is, what sorts of psychological gratifications they derive from it, or how they employ it to gain or maintain status within their circles of friends. One is thus led to the further conclusion that a deeper probing of these levels of media behaviour might have revealed even more striking differences between those ranged at different points along the delinquency continuum. The mass media, after all, are by their very nature fairly ubiquitous: the same material is available to most people. It is at the level of differential perception, memory, usage and meaning that we may expect to find differences which are related to social structural and social psychological variables. And this argument, in its turn, tends again to suggest that Pfuhl's major hypothesis ought to be accepted: that there was, in his Washington sample, a significant positive relationship between delinquency and key aspects of mass media behaviour.

In his summary chapter, Pfuhl notes several limitations of the larger study within which he conducted his own researches, and some of these should be mentioned. In the first place, the research was carried out in three small towns, so that no "large, urban, metropolitan youth are included". In other words, hard-core delinquents, or those who had fully accepted the values of a delinquent subculture, were less likely to have been included in the sample. In its turn, of course (though Pfuhl himself does not advance the argument), this may have acted to weaken the correlations between delinquency and mass media exposure and usage, since the delinquents in the sample were likely to be themselves somewhat 'weak' in their delinquent orientation. It is likely that this characteristic of the sampling frame was aggravated by the fact that high school 'drop-outs' were by definition excluded from the sample, while it is just among this group that we should expect to find the majority of the older delinquents. Unfortunately, Pfuhl does not provide any figures on the rates of drop-out for the schools used in the research, so that it is impossible to guess what sort of effect this may have had. Again, however, the use of school samples may have meant that fewer actual delinquents were included than might otherwise have been the case, thus further weakening the discovered relationships. In fairness to Pfuhl himself, it should be said that he was not responsible for the sampling plan used, though one must also remember that the entire investigation was directed to establishing the social concomitants of juvenile delinquency.

All social research involves compromises, and this is certainly true of large-scale survey research conducted within a highly structured administrative framework. In the case of Pfuhl's research, as noted above, the media indices can be criticized for their superficiality. At the same time, five different media were studied (cinema, television, radio, comics and novels), so that lack of depth is somewhat compensated for by considerable breadth. In the study to be reported below, attention was confined to one medium, television. The two other pieces of published research which best exemplify the straightforward survey or correlational approach to the suggested relationship between mass media and juvenile delinquency are each also concerned with only one medium.

In Hoult's study, two groups of adolescents were matched in terms of sex, age, level in school and socio-economic status, and the behaviour investigated was the reading of comic books.[6] The delinquents claimed to read twice as many comic books as the non-delinquents, and the difference was mainly due to the greater numbers of 'harmful' and 'questionable' comics read by the delinquents:

there was far less difference in the reading of Hoult's third type, 'harmless' comics.

Berninghausen and Faunce used members of an institution for male delinquents in their study of the reading of sensational books.[7] A control group of non-delinquent boys was recruited, with one-to-one matching on the basis of age, level of school, type of school, intelligence, reading ability and father's occupation. In this case, no differences between the two groups were discovered. However, the delinquents' reading ability was considerably below that which would have been predicted on the basis of their intelligence.

In both of these latter investigations, as in the present study, the control group of non-delinquents played a vital part. Without the inclusion of such a group in the research design, it would be impossible to say whether the media habits of the delinquents studied are uniquely characteristic of delinquents, or merely reflect the tastes and activities of all boys or girls of a particular age and type of social background. The question of the matching of control groups will be returned to in the description of the present study.

On the assumption that social research and sociological theory ought ideally to be cumulative, we can at this point ask whether the results of the three studies so far described are similar, and if so, what general conclusions and further hypotheses can be drawn from them. Although there are differences between the samples employed, it is worth noting that both Pfuhl and Hoult would tend to support a link between delinquency and the reading of crime comics amongst girls, although for boys the relationship does not hold in Pfuhl's research. Again, while Pfuhl found that delinquent boys were more interested in crime and mystery novels than non-delinquents, Berninghausen and Faunce arrived at just the opposite conclusion. In general, though, we might say that a link between delinquency and certain aspects of media behaviour had been thrown up often enough for this sort of relationship to be worth looking for under other circumstances, though the correlation is probably not strong enough to emerge whatever the other situational factors involved. Of course, these comparisons can only be made amongst the three studies in terms of degrees of exposure: psychological usage (the exploration of which may itself yield clues about causal probabilities) remains an uncharted level of investigation in all three cases. Although the study to be reported below was not confined to an exploration of differences in psychological usage, it proved that the analysis of the data became most fruitful when tentative arguments about such differences could be advanced. But the administrative framework of the research must first be described.

II

THE PRESENT RESEARCH

Design of the study

In the present investigation, the sample of delinquents was composed of all male and female adolescents between the ages of 10 and 20 who had been placed on probation by the juvenile courts during the months of February, March, and April, 1966, in the Leicestershire, Nottinghamshire, Staffordshire, and Stoke-on-Trent areas.

Of course, taking probationers as the delinquent sample means accepting a purely legal definition of delinquency. Some would argue that this sort of definition has limited sociological relevance. It is no doubt true that many cases of deviant adolescent behaviour go unnoticed or unpunished, and that those dealt with by the juvenile courts represent only the tip of an iceberg. Yet the legally-formulated grounds on which a particular activity is categorized as delinquent do have their roots in widely diffused social values. Although the researcher might ideally prefer to study a sample of *all* those young people who have engaged in socially deviant behaviour of a relatively serious sort, the administratively more convenient decision to study only those placed on probation need not seriously invalidate his work. However, recidivists and those who have committed more serious offences will tend to be under-represented, since some delinquents of both these kinds will have been sent to Approved School or Borstal, or to a Detention Centre.

The sample of adolescents designed to provide a direct comparison with the probationer sample, and referred to throughout as the control sample, was drawn from primary and secondary schools in Leicestershire, and from the membership of youth clubs in the city of Leicester. In order to show whether delinquent status in itself made any difference to an individual's pattern of television behaviour, it was necessary to draw a control sample as similar as possible to the probationer sample in all respects *other than* delinquent status. Matching was to be carried out in terms of age, sex, socio-economic status, intelligence, and school attainment. At the end of this chapter we make some additional comments about the meaningfulness of this type of matched-sample design.

It was decided to include in the control sample one quarter as many boys at each age level as were likely to be included in the probationer sample. Since it was known that few girls are placed on probation, it was decided to recruit as many girls into the control sample as were likely to be included in the probationer sample, except for the oldest age group (those between 17 and 20), where the control sample would include only half as many girls as the probationer sample. The decision to make this latter arrangement sprang from the purely practical problem of finding a sufficient number of suitable older girls in youth clubs.

To provide a further series of comparisons with both the probationer sample and the control sample, a third group of adolescents was also recruited for interview. The age and sex composition of this third sample was designed to be similar to that of the two other samples, but it was to include adolescents of higher socio-economic status, higher intelligence and better school attainment.

Since it was known that both socio-economic status and intelligence are likely to affect the amount and type of exposure to television, the inclusion in the study design of this third sample made it possible to determine whether or not the questions asked of the adolescents were sensitive enough to differentiate between groups which were *known* to be likely to differ in their television behaviour. If a particular question produced data which indicated that there was a difference between this third sample and the control sample, but not between the control sample and the probationer sample, then this would provide rather more certain evidence that the lack of difference between the two latter groups indicated a genuine similarity, and was not merely the result of an insensitive question. On the basis of the general socio-economic status of this third sample, it will be referred to below as the lower middle-class sample. The research design called for this sample to be of the same size as the control sample. Care was of course taken to make sure that neither the control sample nor the lower middle-class sample contained any individuals who had ever been on probation or 'in trouble' in some less formal way.

Taking into account the desired size relationships between the control and lower middle-class samples and the probationer sample, the numbers of adolescents needed to fill the quota requirements set for the two former groups were calculated on the basis of the published criminal statistics for 1963. The Principal Probation Officers in the four areas from which the probationer sample was to be recruited also supplied estimates of the numbers of new probation cases likely to occur in their areas during the stipulated period. After certain adjustments had been made, 334 probationers were finally included in the study, while the total sizes of the control and lower middle-class samples were 144 and 185 respectively.

Appendix C gives a full description of how the numbers of respondents needed for the two non-probationer samples were arrived at. Appendix C also deals with the problem of the representativeness of the probationer sample. We had originally hoped to gather data about all members of the universe defined at the beginning of this section. Unfortunately, this was not possible.

All the members of the probationer sample were interviewed individually by the Probation Officers to whom they had been assigned. The research staff also had access to the background information about each probationer which is collected as a matter of course by the Probation Service. Individual interviews were also carried out with all members of the two other samples. Interviews took place either at the schools or in the youth clubs. Certain additional, background information was also collected from school records, teachers, headmasters, and youth club leaders.

The interview schedule used to collect data on respondents' television-viewing behaviour is set out in Appendix A, and this schedule was of course common to all three samples. It should, however, be pointed out that different schedules were used to collect background data from the probationer sample and the two non-probationer samples. Background data on probationers was drawn from the Record of Supervision (Part A) form normally employed by the Probation Service. A comparable schedule was developed for use with the control and lower middle-class samples, but the format of some questions was different from that used on the Part A form. Thus there were some differences in the way certain items of background information were collected from the probationer and non-probationer samples. The relevant schedules are set out in Appendix B.

The survey was in general of a purely exploratory nature, and no specific research hypotheses were set up. However, an effort was made to collect data about as many aspects of television-viewing behaviour as possible. Certain findings from the survey are tentatively explained below in terms of a number of relatively well-established sociological theories of juvenile delinquency. It should, however, be stressed that interpretation constitutes here an inherently *ex post facto* exercise, and one that is, therefore, less than entirely satisfactory. Further research would be needed to test the set of hypotheses suggested by the data reported in subsequent sections.

This study could only have been carried out with that high degree of co-operation on the part of Probation Officers, teachers and youth club leaders which was almost universally achieved. The fact that many of the data were thus collected by remote control did mean, however, that a variety of gaps still remained at the end of the field-work period. For this reason, a number of the comparisons shown in the tables included below are based on fewer respondents than the total numbers of cases finally included in the various samples. The effective sample sizes are indicated either by the 'Total' line at the foot of a particular table, or by a figure in parentheses. Where no such figure appears, it may be assumed that the numbers of cases entering into the table are as shown in Table 1.

Table 1

The three samples broken down by sex and age groups

Age group	*Probationer sample*	*Control sample*	*Lower M.C. sample*
	MALES		
10 - 12	60 (21%)	24 (24%)	26 (20%)
13 - 14	98 (35%)	32 (33%)	33 (26%)
15 - 16	72 (26%)	23 (23%)	35 (28%)
17 - 20	51 (18%)	19 (19%)	33 (26%)
Total	281 (100%)*	98 (100%)	127 (100%)

Probationer sample and Control sample: $\chi^2 = 0.62$; N.S.
Probationer sample and Lower M.C. sample: $\chi^2 = 4.92$; N.S.
Control sample and Lower M.C. sample: $\chi^2 = 2.65$; N.S.

Age group	*Probationer sample*	*Control sample*	*Lower M.C. sample*
	FEMALES		
10 - 12	3 (6%)	9 (20%)	10 (17%)
13 - 14	17 (32%)	16 (35%)	16 (28%)
15 - 16	11 (21%)	13 (28%)	17 (29%)
17 - 20	22 (42%)	8 (17%)	15 (26%)
Total	53 (100%)	46 (100%)	58 (100%)

Probationer sample and Control sample: $\chi^2 = 9.07$; $p < 0.05$
Probationer sample and Lower M.C. sample: $\chi^2 = 6.19$; N.S.
Control sample and Lower M.C. sample: $\chi^2 = 1.37$; N.S.

* The convention of figures rounding to 100% has been used throughout.

No explicit hypotheses were set up prior to the analysis, and two-tailed statistical tests have therefore been used throughout. Where data were in the form of frequencies, chi-squared has been used to test the statistical significance of inter-sample differences. In some tables, the categories had to be collapsed to yield expected cell values of five or more. The relevant degrees of freedom are indicated only in these cases. Where it seemed reasonable to assume an interval level of measurement (as with the number of hours of television viewed per week), *t*-tests have been carried out for each of the three *pairs* of samples for each sex. Attribute data have been analysed by means of difference-of-percentages tests. The boys' samples were all quite large, so it must be remembered that socially insignificant differences may have resulted in statistical significance at well beyond the five per cent level in some cases. In the interview, respondents were often asked parallel sets of questions (for example, they were asked to name up to six favourite television programmes). In a number of tables, data from the various parts of such questions have been conflated, so that the numbers in the 'Total' line or in brackets are much greater than those given in Table 1. In such tables, statistically significant differences should be treated with caution, since the numbers of 'cases' are artificially inflated.

We are aware of the criticisms commonly made of those who apply inductive statistics to non-random samples. However, we feel that such statistics here furnish a useful heuristic device.

As Table 1 makes plain, there was only a relatively small number of girls in each of the three samples. In addition, the degree of matching achieved between the girls in the probationer and control samples was not very satisfactory. The weaknesses involved are explored more fully below. These two factors mean that we cannot place very great faith in the validity of the girls' data. However, two things have motivated us to include these findings. In the first place, an overall comparison between boys and girls often proved useful in suggesting interpretations of boys' data and in accounting for inter-sample differences among the boys. Secondly, particularly so far as our data bear on 'pop' music, the girls' responses seemed particularly interesting in themselves, while the inter-sample differences are so striking at this point that we believe they would still have emerged from the analysis of more satisfactory data. However, much greater attention is paid to the boys' data, and most of the interpretative hypotheses have to do with boys and the male adolescent role. Statistical tests have in general not been applied in making inter-sex comparisons.

Demographic comparisons between samples

So far as male respondents were concerned, the research design called for the three samples to be matched in terms of their age distributions. Table 1 shows that there were in fact no statistically significant differences between the age distributions of the three samples of boys, when each sample is broken down into four age groups. In the case of girls, efforts were made to recruit into the two non-probationer samples as many respondents as there were likely to be in the probationer sample, except in the case of the 17- to 20-year-old age group. Here the target set was half the number of girls of this age likely to be included amongst the probationers. The table shows that this policy did result in the probationer sample containing more older girls than the two other samples, giving a significant difference in age distribution between the probationer and control samples. Caution therefore needs to be exercised in interpreting differences between these two samples when the aspect of television behaviour under consideration is known to be affected by age, or thought likely to be so affected. No such caution is required in interpreting differences between the three groups of boys.

The research design also called for the matching of the three samples in terms of socio-economic status, as well as age. The data collected during the course of the survey provided a number of indices of social class, and it is therefore also possible to examine whether this aspect of matching between samples was achieved. In fact, as Table 2 indicates, there were significant differences for both males and females between the probationer sample and the control sample

Table 2
Fathers' occupations

Category	*Probationer sample*	*Control sample*	*Lower M.C. sample*
	MALES		
Executive, managerial or supervisory	17 (7%)	3 (3%)	35 (29%)
Skilled manual	45 (19%)	30 (33%)	35 (29%)
Semi-skilled manual	119 (50%)	37 (40%)	28 (23%)
Unskilled manual	50 (21%)	10 (11%)	6 (5%)
Other	7 (3%)	12 (13%)	16 (13%)
Total	238 (100%)	92 (100%)	120 (100%)
Probationer sample and Control sample:		$\chi^2 = 23.76$; $p < 0.001$	
Probationer sample and Lower M.C. sample:		$\chi^2 = 70.68$; $p < 0.001$	
Control sample and Lower M.C. sample:		$\chi^2 = 26.92$; $p < 0.001$	
	FEMALES		
Executive, managerial or supervisory	2 (5%)	5 (13%)	24 (43%)
Skilled manual	4 (10%)	12 (30%)	16 (29%)
Semi-skilled manual	24 (60%)	17 (43%)	6 (11%)
Unskilled manual	9 (23%)	2 (5%)	4 (7%)
Other	1 (3%)	4 (10%)	6 (11%)
Total	40 (100%)	40 (100%)	56 (100%)
Probationer sample and Control sample:		$\chi^2 = 7.46$; df = 2; $p < 0.05$	
Probationer sample and Lower M.C. sample:		$\chi^2 = 34.91$; df = 3; $p < 0.001$	
Control sample and Lower M.C. sample:		$\chi^2 = 17.09$; df = 3; $p < 0.001$	

in terms of the distributions of fathers' occupations. In both cases, the control sample contained a larger proportion of fathers in skilled manual trades, and a smaller proportion of fathers in semi-skilled and unskilled trades. However, despite these differences, occupations which could be called middle-class account for an almost equally small percentage of cases in both samples. Both the probationer and control samples are overwhelmingly working class. The justification in labelling the lower middle-class sample as such is seen in the far greater proportion of fathers in this sample who were in executive, managerial or supervisory jobs. In the main, the occupations fell into the two latter categories.

Another, less direct index of respondents' socio-economic status was provided by the type of housing their families occupied. According to this measure, as Table 3 shows, there was no difference between the probationers and the control sample, while the houses in which lower middle-class respondents lived were far more likely to be detached or semi-detached than was true of either of the two other samples. A very similar picture emerges from the data collected on house ownership, and presented in Table 4. Members of the lower middle-class sample were roughly four times as likely to live in

Table 3
Housing

House type	*Probationer sample*	*Control sample*	*Lower M.C. sample*
	MALES		
Detached	9 (3%)	2 (2%)	15 (12%)
Semi-detached	38 (14%)	24 (24%)	42 (33%)
Terrace	59 (21%)	24 (24%)	29 (23%)
Council	163 (58%)	44 (45%)	27 (21%)
Other	10 (4%)	4 (4%)	14 (11%)
Total	279 (100%)	98 (100%)	127 (100%)

Probationer sample and Control sample: $\chi^2 = 5.98$; df = 2; N.S.
Probationer sample and Lower M.C. sample: $\chi^2 = 61.67$; $p < 0.001$
Control sample and Lower M.C. sample: $\chi^2 = 21.57$; $p < 0.001$

House type	*Probationer sample*	*Control sample*	*Lower M.C. sample*
	FEMALES		
Detached, semi-detached	10 (19%)	8 (17%)	27 (47%)
Terrace	11 (21%)	12 (26%)	18 (31%)
Council	28 (53%)	23 (50%)	9 (16%)
Other	4 (8%)	3 (7%)	4 (7%)
Total	53 (100%)	46 (100%)	58 (100%)

Probationer sample and Control sample: $\chi^2 = 0.39$; df = 2; N.S.
Probationer sample and Lower M.C. sample: $\chi^2 = 17.33$; df = 2; $p < 0.001$
Control sample and Lower M.C. sample: $\chi^2 = 14.66$; df = 2; $p < 0.001$

Table 4

House ownership

Ownership of house	*Probationer sample*	*Control sample*	*Lower M.C. sample*
	MALES		
Yes	50 (18%)	12 (13%)	62 (50%)
No	224 (82%)	84 (88%)	63 (50%)
Total	274 (100%)	96 (100%)	125 (100%)

Probationer sample and Control sample: $\chi^2 = 1.30$; N.S.
Probationer sample and Lower M.C. sample: $\chi^2 = 40.25$; $p < 0.001$
Control sample and Lower M.C. sample: $\chi^2 = 31.91$; $p < 0.001$

	FEMALES		
Yes	6 (11%)	4 (9%)	31 (54%)
No	47 (89%)	40 (91%)	26 (46%)
Total	53 (100%)	44 (100%)	57 (100%)

Probationer sample and Control sample: χ^2 is indeterminate
Probationer sample and Lower M.C. sample: $\chi^2 = 20.93$; $p < 0.001$
Control sample and Lower M.C. sample: $\chi^2 = 20.54$; $p < 0.001$

Table 5

Intelligence quotients

Probationer sample	*Control sample*	*Lower M.C. sample*
	MALES	
90.2 (N = 202)	88.1 (N = 77)	104.0 (N = 104)

Probationer sample and Control sample: $t = 1.28$; N.S.
Probationer sample and Lower M.C. sample: $t = 9.23$; $p < 0.001$
Control sample and Lower M.C. sample: $t = 9.00$; $p < 0.001$

	FEMALES	
92.1 (N = 36)	86.4 (N = 39)	103.0 (N = 47)

Probationer sample and Control sample: $t = 2.22$; $p < 0.05$
Probationer sample and Lower M.C. sample: $t = 4.65$; $p < 0.001$
Control sample and Lower M.C. sample: $t = 8.45$; $p < 0.001$

houses owned by their parents as was the case with the two other samples. Using housing as an index of social class, therefore, it can be claimed that the probationer and control samples were successfully matched. This is also the case in terms of fathers' occupations if this is taken in a simple working-class/middle-class sense. However, skilled manual workers were over-represented amongst the fathers of control-sample members if one adopts a stricter approach to matching. This question is taken up again later on, when we turn to consider the different working-class subcultures represented by the probationer and control samples.

It was intended that the three samples should be further matched in terms of intelligence, since a considerable body of prior research had indicated that television programme tastes, for example, are correlated with mental ability. Table 5 shows that male probationers did not differ from boys in the control sample in terms of measured intelligence, though female probationers were in fact significantly *more* intelligent on average than members of the control sample. It will be recalled that female probationers were also on average slightly older than their matched controls, so that the discovery of an additional difference means that there are considerably greater grounds for treating comparisons between these groups with some caution. In the case of both sexes, lower middle-class respondents were significantly higher in mean intelligence than the members of the two other samples. Thus in this respect the quota differences between the samples were successfully achieved.

As an indirect check on how well respondents had realized their intellectual abilities within the formal educational system, data on each individual's position within the streaming system of his or her own school was also collected. Table 6 shows that a higher proportion of probationers than of control-sample members were in the top half

Table 6

Positions in schools' streaming systems

Streaming position	*Probationer sample*	*Control sample*	*Lower M.C. sample*
	MALES		
Top half	48 (30%)	13 (15%)	91 (81%)
Bottom half	111 (70%)	76 (85%)	22 (19%)
Total	159 (100%)	89 (100%)	113 (100%)

Probationer sample and Control sample: $\chi^2 = 6.65$; $p < 0.01$
Probationer sample and Lower M.C. sample: $\chi^2 = 65.00$; $p < 0.001$
Control sample and Lower M.C. sample: $\chi^2 = 84.00$; $p < 0.001$

Streaming position	*Probationer sample*	*Control sample*	*Lower M.C. sample*
	FEMALES		
Top half	10 (37%)	7 (18%)	39 (83%)
Bottom half	17 (63%)	32 (82%)	8 (17%)
Total	27 (100%)	39 (100%)	47 (100%)

Probationer sample and Control sample: $\chi^2 = 2.12$; N.S.
Probationer sample and Lower M.C. sample: $\chi^2 = 14.19$; $p < 0.001$
Control sample and Lower M.C. sample: $\chi^2 = 33.67$; $p < 0.001$

Table 7
School records

Index	*Probationer sample*	*Control sample*	*Lower M.C. sample*
MALES			
*Attainment**	2.21	2.45	3.25
Probationer sample and Control sample:			$t = 1.84$; N.S.
Probationer sample and Lower M.C. sample:			$t = 7.94$; $p < 0.001$
Control sample and Lower M.C. sample:			$t = 4.89$; $p < 0.001$
*Conduct**	2.84 (N = 266)	2.56	3.60
Probationer sample and Control sample:			$t = 2.49$; $p < 0.05$
Probationer sample and Lower M.C. sample:			$t = 7.69$; $p < 0.001$
Control sample and Lower M.C. sample:			$t = 8.53$; $p < 0.001$
Absence†	47.93 (N = 136)	49.52 (N = 75)	24.31 (N = 93)
Probationer sample and Control sample:			$t = 0.23$; N.S.
Probationer sample and Lower M.C. sample:			$t = 3.78$; $p < 0.001$
Control sample and Lower M.C. sample:			$t = 3.67$; $p < 0.001$
FEMALES			
*Attainment**	2.13	2.11	3.43
Probationer sample and Control sample:			$t = 0.08$; N.S.
Probationer sample and Lower M.C. sample:			$t = 4.97$; $p < 0.001$
Control sample and Lower M.C. sample:			$t = 5.61$; $p < 0.001$
*Conduct**	2.94 (N = 47)	2.87 (N = 45)	3.40 (N = 55)
Probationer sample and Control sample:			$t = 0.40$; N.S.
Probationer sample and Lower M.C. sample:			$t = 2.93$; $p < 0.01$
Control sample and Lower M.C. sample:			$t = 4.09$; $p < 0.001$
Absence†	62.29 (N = 24)	31.37 (N = 35)	14.98 (N = 40)
Probationer sample and Control sample:			$t = 2.69$; $p < 0.05$
Probationer sample and Lower M.C. sample:			$t = 5.07$; $p < 0.001$
Control sample and Lower M.C. sample:			$t = 3.11$; $p < 0.01$

* The higher the score, the 'better' the performance.
† Number of half-day absences during previous (or last) school year.

of their respective schools, and that the difference was statistically significant in the case of boys. While this situation may reflect the fact that the probationers were marginally more intelligent on average than the members of the control sample, it is still somewhat surprising. Tables 5 and 6 taken in conjunction certainly do *not* suggest that probationers were failing to do as well at school as their abilities would lead one to expect. Indeed, the tendency seems to be somewhat in the other direction. However, other data paint a quite different picture.

Adult judges—mainly class teachers or head teachers—were asked to rate each respondent on various 5-point scales. In the first instance, a rating was requested in terms of the boy's or girl's educational attainment in relation to the class or stream he or she was in at the time of the survey, or had been a member of at the time of leaving school. A score of 5 was assigned for the rating 'well above average', down to a score of 0 for a rating of 'well below average'. The average scores for each group on this scale are shown in Table 7, where it is seen that neither male nor female probationers did significantly less well than was the case with control-sample members. Thus male probationers, besides their intelligence and their streaming position in their schools, were also not 'under achievers' when compared with the matched controls. The fact that girls did not differ significantly may be a simple function of the female probationers being significantly more intelligent on average than their matched controls. Alternatively, it may reflect a more important difference between probationers of the two sexes.

Teachers also rated respondents on a similar 5-point scale in terms of their school conduct. Table 7 shows that here the male probationers scored higher on average than did the adolescents in the control sample, but that the difference was not statistically significant for girls. Both boys and girls in the middle-class sample were rated as significantly superior in this respect to their peers in the two other samples. Thus only in terms of school conduct did male probationers do considerably better than one might have been led to expect on the basis of their ability and placement in the school. However, it has to be pointed out that some biasses may have been operating when teachers rated their pupils on these scales.

Since teachers would in most cases have known that the members of the primary sample were juvenile delinquents—in most cases the teachers were approached directly by the Probation Officer concerned—they can be expected to have allowed this knowledge to colour their assessments to some extent, with the probable result that the probationers received somewhat higher ratings than their actual performance in school merited. Rather similarly, since it was the school, in the person of the head or class teacher, which was responsible for the initial assignment of pupils to the control or lower middle-class sample, when the time came to make further assessments of the selected pupils there was no doubt a tendency to validate earlier judgments by giving appropriate ratings. Some question must therefore be attached to the validity of these measures of school attainment and conduct.

No such doubts attach to the objective data on absences from school. These are also shown in Table 7, but do not suggest the expected relationships between the three samples. Only female probationers were absent from school significantly more often on average than was true of members of the control sample. However, both boys and girls in the latter group had been absent significantly more often than members of the lower middle-class sample. The lack of any difference between the male probationers and their matched controls suggests that not only were the probationers at least as intelligent as the control group, they were also as strongly committed to the values of the school system, whatever may be the theoretical explanation of this somewhat surprising finding in terms of subcultural differences. In fact, these findings taken by themselves might lead one to argue against the appropriateness of any subcultural theory of delinquency, since such a theory posits opposition to all middle-class institutions, particularly perhaps the school. However, we shall argue later on that the male probationers *were* probably opposed to the school and what it stands for, despite these data on absences.

Besides some differences in within-school behaviour, the proba-

Table 8
Fathers

Status	*Probationer sample*	*Control sample*	*Lower M.C. sample*
	MALES		
Living with respondent	227 (83%)	88 (90%)	116 (94%)
Not living with respondent	31 (11%)	3 (3%)	3 (2%)
Dead	14 (5%)	7 (7%)	4 (3%)
Total	272 (100%)	98 (100%)	123 (100%)

Probationer sample and Control sample: $\chi^2 = 6.29$; $p < 0.05$
Probationer sample and Lower M.C. sample: $\chi^2 = 9.71$; $p < 0.01$
Control sample and Lower M.C. sample: $\chi^2 = 0.95$; df = 1; N.S.

Status	*Probationer sample*	*Control sample*	*Lower M.C. sample*
	FEMALES		
Living with respondent	38 (78%)	40 (87%)	56 (97%)
Not living with respondent	9 (18%)	5 (11%)	1 (2%)
Dead	2 (4%)	1 (2%)	1 (2%)
Total	49 (100%)	46 (100%)	58 (100%)

Probationer sample and Control sample: $\chi^2 = 0.86$; df = 1; N.S.
Probationer sample and Lower M.C. sample: $\chi^2 = 7.29$; df = 1; $p < 0.01$
Control sample and Lower M.C. sample: χ^2 is indeterminate

Table 9

Mothers

Status	*Probationer sample*	*Control sample*	*Lower M.C. sample*
	MALES		
Living with respondent	243 (87%)	92 (95%)	125 (98%)
Not living with respondent	24 (9%)	1 (1%)	0 (0%)
Dead	11 (4%)	4 (4%)	2 (2%)
Total	278 (100%)	97 (100%)	127 (100%)

Probationer sample and Control sample: $\chi^2 = 3.43$; df = 1; N.S.
Probationer sample and Lower M.C. sample: $\chi^2 = 11.45$; df = 1; $p < 0.001$
Control sample and Lower M.C. sample: χ^2 is indeterminate

Status	*Probationer sample*	*Control sample*	*Lower M.C. sample*
	FEMALES		
Living with respondent	44 (83%)	42 (91%)	56 (97%)
Not living with respondent	7 (13%)	2 (4%)	1 (2%)
Dead	1 (2%)	2 (4%)	1 (2%)
Total	52 (100%)	46 (100%)	58 (100%)

Probationer sample and Control sample: $\chi^2 = 0.48$; df = 1; N.S.
Probationer sample and Lower M.C. sample: χ^2 is indeterminate
Control sample and Lower M.C. sample: χ^2 is indeterminate

Table 10

Fathers' work status

Category	*Probationer sample*	*Control sample*	*Lower M.C. sample*
	MALES		
Regularly employed	146 (74%)	84 (91%)	116 (97%)
Unemployed	52 (26%)	8 (9%)	3 (3%)
Total	198 (100%)	92 (100%)	119 (100%)

Probationer sample and Control sample: $\chi^2 = 10.77$; $p < 0.01$
Probationer sample and Lower M.C. sample: $\chi^2 = 27.58$; $p < 0.001$
Control sample and Lower M.C. sample: χ^2 is indeterminate

Category	*Probationer sample*	*Control sample*	*Lower M.C. sample*
	FEMALES		
Regularly employed	30 (86%)	39 (93%)	56 (100%)
Unemployed	5 (14%)	3 (7%)	0 (0%)
Total	35 (100%)	42 (100%)	56 (100%)

Probationer sample and Control sample: χ^2 is indeterminate
Probationer sample and Lower M.C. sample: χ^2 is indeterminate
Control sample and Lower M.C. sample: χ^2 is indeterminate

tioners were also differentiated from the controls by reason of their domestic circumstances. Of course, no attempt had been made to match the samples in this area. In the case of both sexes, significantly more probationers were *not* living with their natural fathers than was the case with the control sample, as Table 8 shows. Just the same held for respondents' mothers, as seen in Table 9. A number of studies of the social factors which predict juvenile delinquency have of course shown that delinquents are more than usually likely to come from broken homes of one sort or another, so these data occasion no surprise.[1]

The closely related notion that delinquents tend to lack a father who provides an adequate male identification figure is also supported indirectly by data from the survey. Table 10 shows that the fathers of both male and female probationers were less likely to be in regular employment than the fathers of members of the control sample, and the difference reached a level of statistical significance in the case of the males. Probationers' fathers were also less likely to be in good health than was true of other groups, with a particularly striking (and significant) difference in the case of boys. The relevant findings are shown in Table 11. However, the mothers of probationers, too, were in less good health than was the case with the other two samples, with the difference being very significant statistically for both sexes. The relevant percentages appear in Table 12. So far as

Table 11
Fathers' health

Health	*Probationer sample*	*Control sample*	*Lower M.C. sample*
	MALES		
Good	170 (66%)	83 (88%)	114 (94%)
Poor	87 (34%)	11 (12%)	7 (6%)
Total	257 (100%)	94 (100%)	121 (100%)

Probationer sample and Control sample: $\chi^2 = 15.70$; $p < 0.001$
Probationer sample and Lower M.C. sample: $\chi^2 = 33.20$; $p < 0.001$
Control sample and Lower M.C. sample: $\chi^2 = 1.70$; N.S.

Health	*Probationer sample*	*Control sample*	*Lower M.C. sample*
	FEMALES		
Good	35 (73%)	34 (79%)	52 (93%)
Poor	13 (27%)	9 (21%)	4 (7%)
Total	48 (100%)	43 (100%)	56 (100%)

Probationer sample and Control sample: $\chi^2 = 0.19$; N.S.
Probationer sample and Lower M.C. sample: $\chi^2 = 6.13$; $p < 0.05$
Control sample and Lower M.C. sample: $\chi^2 = 2.93$; N.S.

Table 12
Mothers' health

Health	Probationer sample	Control sample	Lower M.C. sample
	MALES		
Good	168 (64%)	76 (83%)	107 (85%)
Poor	95 (36%)	16 (17%)	19 (15%)
Total	263 (100%)	92 (100%)	126 (100%)

Probationer sample and Control sample: $\chi^2 = 10.27$; $p < 0.01$
Probationer sample and Lower M.C. sample: $\chi^2 = 17.20$; $p < 0.001$
Control sample and Lower M.C. sample: $\chi^2 = 0.07$; N.S.

Health	Probationer sample	Control sample	Lower M.C. sample
	FEMALES		
Good	25 (52%)	35 (83%)	46 (82%)
Poor	23 (48%)	7 (17%)	10 (18%)
Total	48 (100%)	42 (100%)	56 (100%)

Probationer sample and Control sample: $\chi^2 = 8.49$; $p < 0.01$
Probationer sample and Lower M.C. sample: $\chi^2 = 9.44$; $p < 0.01$
Control sample and Lower M.C. sample: $\chi^2 = 0.01$; N.S.

Table 13
Respondents' health

Health	Probationer sample	Control sample	Lower M.C. sample
	MALES		
Good	244 (87%)	86 (90%)	112 (89%)
Poor	36 (13%)	10 (10%)	14 (11%)
Total	280 (100%)	96 (100%)	126 (100%)

Probationer sample and Control sample: $\chi^2 = 0.20$; N.S.
Probationer sample and Lower M.C. sample: $\chi^2 = 0.11$; N.S.
Control sample and Lower M.C. sample: $\chi^2 = 0.00$; N.S.

Health	Probationer sample	Control sample	Lower M.C. sample
	FEMALES		
Good	46 (87%)	40 (89%)	57 (100%)
Poor	7 (13%)	5 (11%)	0 (0%)
Total	53 (100%)	45 (100%)	57 (100%)

Probationer sample and Control sample: $\chi^2 = 0.00$; N.S.
Probationer sample and Lower M.C. sample: χ^2 is indeterminate
Control sample and Lower M.C. sample: χ^2 is indeterminate

the respondents' own health was concerned, there were no significant differences between the three samples (Table 13).

Given these data on the frequent absence, ill health or irregular employment of probationers' fathers, on the roughly equally frequent

Table 14

Mothers' employment

Category	*Probationer sample*	*Control sample*	*Lower M.C. sample*
	MALES		
Employed full time	45 (18%)	23 (25%)	35 (28%)
Employed part time	60 (23%)	37 (41%)	41 (33%)
Not employed	151 (59%)	31 (34%)	49 (39%)
Total	256 (100%)	91 (100%)	125 (100%)

Probationer sample and Control sample: $\chi^2 = 17.10$; $p < 0.001$
Probationer sample and Lower M.C. sample: $\chi^2 = 13.39$; $p < 0.01$
Control sample and Lower M.C. sample: $\chi^2 = 1.42$; N.S.

Category	*Probationer sample*	*Control sample*	*Lower M.C. sample*
	FEMALES		
Employed full time	12 (26%)	10 (23%)	15 (27%)
Employed part time	7 (15%)	16 (36%)	21 (38%)
Not employed	27 (59%)	18 (41%)	20 (36%)
Total	46 (100%)	44 (100%)	56 (100%)

Probationer sample and Control sample: $\chi^2 = 5.47$; N.S.
Probationer sample and Lower M.C. sample: $\chi^2 = 7.47$; $p < 0.05$
Control sample and Lower M.C. sample: $\chi^2 = 0.34$; N.S.

absence of their mothers, and on their mothers' ill health, it is possible to suggest that the probationers' backgrounds were marked by a considerable measure of social disorganization. This kind of disorganization has become well known as part of one common syndrome to which socially deviant behaviour also belongs. Of course, not all the probationers exhibited this pattern. However, this indirect evidence of the incidence of the syndrome will prove of some value below in explaining certain patterns of television-viewing behaviour which appear to have been characteristic of the probationers.

Popular writers have on occasion suggested a link between juvenile delinquency and the employment of mothers, arguing that working mothers are not in a position adequately to supervise their children. However, as Table 14 shows very clearly, probationers' mothers in the present study were significantly *less* likely to be in any form of employment than was true of either of the two other samples. The pattern held for both males and females, though the difference between the probationers and control-sample members is not statistically significant for girls.[2] These findings, of course, are not meant to imply or prove that the probationers studied were not suffering, or had not suffered, from some form of maternal deprivation. Indeed, most of the available research evidence on disorganized working-class families suggests that various types of maternal deprivation are quite likely to occur. However, these are in general probably of a more subtle sort than that likely to be produced by regular absences from the house at a place of work.

Boys in the probationer sample had significantly more brothers and sisters than boys in the control sample, but although the difference among girls was in the same direction, this was not large enough to be significant (Table 15). Again, although probationers' households tended to be larger than those characteristic of the control sample, the differences were not significant, and the same held true for the index of living density. The relevant group means are also shown in Table 15. The average scores for the lower middle-class sample are more or less what were to be expected, with respondents having fewer siblings, living in smaller households, and consequently being less crowded.

Table 15

Households

Characteristic	*Probationer sample*	*Control sample*	*Lower M.C. sample*
	MALES		
Number of siblings	3.78	3.09	1.79
Probationer sample and Control sample:		$t = 2.44$; $p < 0.05$	
Probationer sample and Lower M.C. sample:		$t = 8.46$; $p < 0.001$	
Control sample and Lower M.C. sample:		$t = 5.58$; $p < 0.001$	
Number of household members	6.23	5.71	4.51
Probationer sample and Control sample:		$t = 1.83$; N.S.	
Probationer sample and Lower M.C. sample:		$t = 7.49$; $p < 0.001$	
Control sample and Lower M.C. sample:		**$t = 5.07$; $p < 0.001$**	
*Living density**	1.21	1.20	0.94
Probationer sample and Control sample:		$t = 0.10$; N.S.	
Probationer sample and Lower M.C. sample:		$t = 6.27$; $p < 0.001$	
Control sample and Lower M.C. sample:		$t = 5.08$; $p < 0.001$	
	FEMALES		
Number of siblings	3.79	3.24	2.38
Probationer sample and Control sample:		$t = 0.96$; N.S.	
Probationer sample and Lower M.C. sample:		$t = 2.53$; $p < 0.05$	
Control sample and Lower M.C. sample:		$t = 1.61$; N.S.	
Number of household members	5.91	5.61	4.98
Probationer sample and Control sample:		$t = 0.59$; N.S.	
Probationer sample and Lower M.C. sample:		$t = 2.11$; $p < 0.05$	
Control sample and Lower M.C. sample:		$t = 1.44$; N.S.	
*Living density**	1.18	1.14	0.98
Probationer sample and Control sample:		$t = 0.49$; N.S.	
Probationer sample and Lower M.C. sample:		$t = 2.34$; $p < 0.05$	
Control sample and Lower M.C. sample:		$t = 1.88$; N.S.	

* Persons in household/rooms in house.

Table 16

*Ratings of home background**

Probationer sample	*Control sample*	*Lower M.C. sample*
	MALES	
1.79	1.53	2.51

Probationer sample and Control sample: $t = 2.78$; $p < 0.01$
Probationer sample and Lower M.C. sample: $t = 8.88$; $p < 0.001$
Control sample and Lower M.C. sample: $t = 10.23$; $p < 0.001$

Probationer sample	*Control sample*	*Lower M.C. sample*
	FEMALES	
1.72	1.59	2.53

Probationer sample and Control sample: $t = 0.85$; N.S.
Probationer sample and Lower M.C. sample: $t = 4.99$; $p < 0.001$
Control sample and Lower M.C. sample: $t = 6.61$; $p < 0.001$

* The higher the score, the 'better' the home background.

The general home background of each respondent was rated on a 5-point scale, with a score of 5 given to a rating of 'well above average' in the familiar fashion. Rather surprisingly, both male and female probationers achieved a higher mean rating than did members of the control sample, with the difference being statistically significant in the case of boys. However, this situation may be a reflection of the fact that the ratings of the probationers' backgrounds were provided by persons other than those who provided the comparable ratings for members of the two other samples. These findings appear in Table 16.

Of all these comparisons between the three samples in terms of demographic characteristics or background, that focusing on fathers' occupations is no doubt the most important from the point of view of the original research design. Table 2 shows that there was a statistically and perhaps methodologically significant difference between the class distributions of male respondents in the probationer and control samples. Ideally, it might have been preferable to hold parental occupation constant in the following tables so as to separate out any effect of social status on television behaviour. Unfortunately, this would mean in its turn that the age distribution would not be the same in all samples. Thus while the class difference between the two key groups of respondents must be remembered throughout, it should be stressed that this was a matter of different skill levels *within* the working class, and that these two groups were effectively matched in terms of intelligence (and of course sex).

There is, however, another way to look at the whole set of

differences between the probationer and control samples. So far we have mainly been concerned with the extent to which the matching called for in the research design was actually achieved. In addition, it has been suggested that the probationers came mainly from a disorganized working-class stratum. The comparison implicit in this latter comment can now be brought out more clearly.

People commonly assume that juvenile delinquency is a 'working-class phenomenon', and leave it at that. But several empirical studies suggest that the situation is not so simple. It is not so much working-class adolescents in general who tend to become delinquent, as adolescents from a particular stratum of the working class. For example, in their study of 'Radby' in Nottinghamshire, Jephcott and Carter distinguish between the 'rough' and the 'respectable' working-class families, and show that the former are far more likely to produce delinquents than the latter.[3] Miller makes a very similar distinction between the lower class and the respectable working class.[4] In the light of the data reported above, there is good reason to see the differences between the probationer and control samples in similar terms.

The findings on father's work records and on parents' health are particularly pertinent here. But the sort of work which respondents' fathers did is also relevant. The fact that probationers' fathers were more likely to be in unskilled manual occupations is quite congruent with the notion that this sample was primarily drawn from the 'rough', disorganized, lower working class. Despite the fact that the two groups of boys were matched in terms of intelligence, it may be more realistic, and more meaningful from the point of view of interpretation, to think of the probationer and control samples as being drawn from *distinct* working-class subcultures, rather than as being matched. This kind of contrast did not motivate the original research design, nor is it mentioned here to excuse weaknesses in the degree of matching actually achieved. However, the relevance of using the more usual sorts of demographic matching variables in studies of the present sort is brought into question. Indeed, if one had attempted to locate a control sample matched in terms of *all* the subcultural variables discussed here, it seems quite possible that the attempt would have proved abortive. 'Rough' working-class families with *no* legal record of juvenile delinquency may be few and far between while similar families where adolescent members have shown no *tendency* to become delinquent are no doubt even fewer.

The ideal matched-sample design is often thought of as involving matching on *all* variables other than the chosen independent variable.

(Legal status as a delinquent was thought of as the independent variable in the present study.) Yet such an ideal design may be impossible to achieve, and can indeed be seen as sociologically nonsensical. Thus while reference will be made subsequently to the matching on intelligence achieved for boys in the probationer and control samples, the suggested cultural differences between these samples will be seen to prove of far greater value in the development of an explanatory account of the main findings.

conducting interviews
→ collect as many statements
about it as possible

III

TELEVISION AND TELEVISION VIEWING

Importance of television

A number of questions included in the interview schedule were designed to discover whether there were any differences between samples in the importance they attached to television, or in the salience of the role it played in their lives. Some of these questions were very indirect, a number more pointed.

Right at the beginning of the interview, before mention had been made of the fact that the main focus of the session was to be television, respondents were asked what part of an ordinary weekday they most enjoyed. It seemed possible that the attachment of greater importance to television, the other mass media, and leisure pursuits in general might be reflected in differences between samples in the frequency with which the evening—the non-working part of the day—was referred to. However, Table 17 shows that no such differences emerged from the analysis of the data, although all samples expressed a majority preference for the evening over the morning or afternoon; and this held for both sexes. When asked *why* they enjoyed the particular part of the day they had referred to, a number of interviewees mentioned television or one of the other mass media. But here again there were in general no differences between samples in the proportions mentioning the media, although significantly fewer female probationers made this sort of reference when compared with lower middle-class girls (Table 18).

Towards the end of the interview, respondents were asked whether they would choose the 'pictures' or television, if they had a free choice and there was something they liked equally on both media. As Table 19 shows, although rather more control-sample members of both sexes chose the cinema when compared to the two other samples, none of the differences between samples was significant. Nor, taking all respondents together, did one of the media enjoy a particularly strong advantage over the other.

When asked to give one word to describe the television fare which they actually saw, respondents provided a set of answers—not all of them of one word—which were largely evaluative in tone. Categorization of the responses into positive and negative groups led

Table 17

Part of the day most enjoyed

Part of day	*Probationer sample*	*Control sample*	*Lower M.C. sample*
	MALES		
Morning or afternoon	78 (37%)	36 (45%)	39 (37%)
Evening	131 (63%)	44 (55%)	66 (63%)
Total	209 (100%)	80 (100%)	105 (100%)

Probationer sample and Control sample: $\chi^2 = 1.12$; N.S.
Probationer sample and Lower M.C. sample: $\chi^2 = 0.01$; N.S.
Control sample and Lower M.C. sample: $\chi^2 = 0.86$; N.S.

	FEMALES		
Morning or afternoon	12 (33%)	16 (36%)	14 (28%)
Evening	24 (67%)	28 (64%)	36 (72%)
Total	36 (100%)	44 (100%)	50 (100%)

Probationer sample and Control sample: $\chi^2 = 0.00$; N.S.
Probationer sample and Lower M.C. sample: $\chi^2 = 0.09$; N.S.
Control sample and Lower M.C. sample: $\chi^2 = 0.42$; N.S.

Table 18

Mass media as a reason for enjoying a particular part of the day

Reference to mass media	*Probationer sample*	*Control sample*	*Lower M.C. sample*
	MALES		
Yes	39 (15%)	17 (19%)	17 (13%)
No	220 (85%)	74 (81%)	109 (87%)
Total	259 (100%)	91 (100%)	126 (100%)

Probationer sample and Control sample: $\chi^2 = 0.42$; N.S.
Probationer sample and Lower M.C. sample: $\chi^2 = 0.06$; N.S.
Control sample and Lower M.C. sample: $\chi^2 = 0.72$; N.S.

	FEMALES		
Yes	6 (13%)	13 (30%)	20 (36%)
No	41 (87%)	31 (70%)	36 (64%)
Total	47 (100%)	44 (100%)	56 (100%)

Probationer sample and Control sample: $\chi^2 = 2.92$; N.S.
Probationer sample and Lower M.C. sample: $\chi^2 = 5.97$; $p < 0.05$
Control sample and Lower M.C. sample: $\chi^2 = 0.19$; N.S.

Table 19

Television and the cinema as favourite media

Medium preferred	*Probationer sample*	*Control sample*	*Lower M.C. sample*
	MALES		
Cinema	147 (56%)	57 (63%)	66 (55%)
Television	114 (44%)	34 (37%)	55 (45%)
Total	261 (100%)	91 (100%)	121 (100%)

Probationer sample and Control sample: $\chi^2 = 0.86$; N.S.
Probationer sample and Lower M.C. sample: $\chi^2 = 0.05$; N.S.
Control sample and Lower M.C. sample: $\chi^2 = 1.08$; N.S.

	FEMALES		
Cinema	24 (49%)	23 (58%)	25 (47%)
Television	25 (51%)	17 (43%)	28 (53%)
Total	49 (100%)	40 (100%)	53 (100%)

Probationer sample and Control sample: $\chi^2 = 0.35$; N.S.
Probationer sample and Lower M.C. sample: $\chi^2 = 0.20$; N.S.
Control sample and Lower M.C. sample: $\chi^2 = 0.60$; N.S.

to the findings shown in Table 20. For all samples and for both sexes, there were roughly speaking four positive references for every negative reference. But none of the inter-sample differences was significant.

Table 20

Evaluations of television actually viewed

Evaluation	*Probationer sample*	*Control sample*	*Lower M.C. sample*
	MALES		
Positive	95 (75%)	29 (76%)	35 (63%)
Negative	32 (25%)	9 (24%)	21 (38%)
Total	127 (100%)	38 (100%)	56 (100%)

Probationer sample and Control sample: $\chi^2 = 0.00$; N.S.
Probationer sample and Lower M.C. sample: $\chi^2 = 2.30$; N.S.
Control sample and Lower M.C. sample: $\chi^2 = 1.40$; N.S.

	FEMALES		
Positive	14 (67%)	18 (86%)	11 (73%)
Negative	7 (33%)	3 (14%)	4 (27%)
Total	21 (100%)	21 (100%)	15 (100%)

Probationer sample and Control sample: $\chi^2 = 1.18$; N.S.
Probationer sample and Lower M.C. sample: χ^2 is indeterminate
Control sample and Lower M.C. sample: χ^2 is indeterminate

Table 21

Watching television merely because viewing enjoyed

Probationer sample	*Control sample*		*Lower M.C. sample*
	MALES		
216 (77%)	74 (76%)		86 (68%)
Probationer sample and Control sample:		Z = 0.27; N.S.	
Probationer sample and Lower M.C. sample:		Z = 1.95; N.S.	
Control sample and Lower M.C. sample:		Z = 1.27; N.S.	
	FEMALES		
45 (85%)	40 (87%)		44 (76%)
Probationer sample and Control sample:		Z = 0.29; N.S.	
Probationer sample and Lower M.C. sample:		Z = 1.19; N.S.	
Control sample and Lower M.C. sample:		Z = 1.43; N.S.	

Respondents were directly asked whether or not they ever watched television 'because you just like watching', and some three-quarters of all respondents said that they did sometimes view for this reason. But as the various percentages shown in Table 21 indicate, there were no inter-sample differences, although girls in all three samples said 'yes' to this question more often than boys.

The interview questions from which these data are drawn were all relatively crude. They may have been too insensitive to reveal any slight but actual differences between samples. However, the findings lend no support to any suggestion that television in general had a higher salience for one of the samples as compared with the others.

One current subcultural theory of delinquency suggests that those who, through their being born into the lower working class, or into a particularly deprived group, are unable to achieve middle-class goals by legitimate means, will turn alternatively to illegitimate means, that is, to delinquency. In other words, delinquency arises when opportunities to achieve those material and social goals widely regarded as normal and proper are effectively absent or blocked.[1] Since the advertisements carried by the mass media paint a picture of a world where the acquisition and enjoyment of material goods is a normal goal, it was thought possible that delinquents might manifest a particular orientation to such advertisements. This could conceivably take one of two forms. If delinquents saw the goals represented by the advertisements as really out of reach, then they might tend to devalue the goals and react negatively to the advertisements, according to some rule of cognitive consistency. Alternatively, delinquents might be expected to react particularly positively to advertisements, since they can be seen as representing a fantasy gratification of desires inevitably destined to remain *un*gratified in real life. Thus one might predict that delinquents would show *either* a particularly negative *or* a particularly positive reaction to television advertisements.

Respondents were therefore asked to give a word to describe the commercials shown on the ITV network, and Table 22 shows that all

Table 22

Evaluation of commercials

Type of reaction	*Probationer sample*	*Control sample*	*Lower M.C. sample*
	MALES		
Positive	447 (34%)	182 (38%)	163 (29%)
Negative	868 (66%)	298 (62%)	408 (71%)
Total	1315 (100%)	480 (100%)	571 (100%)

Probationer sample and Control sample: $\chi^2 = 2.21$; N.S.
Probationer sample and Lower M.C. sample: $\chi^2 = 5.15$; $p < 0.05$
Control sample and Lower M.C. sample: $\chi^2 = 9.93$; $p < 0.01$

Type of reaction	*Probationer sample*	*Control sample*	*Lower M.C. sample*
	FEMALES		
Positive	75 (30%)	89 (39%)	87 (34%)
Negative	173 (70%)	139 (61%)	169 (66%)
Total	248 (100%)	228 (100%)	256 (100%)

Probationer sample and Control sample: $\chi^2 = 3.69$; N.S.
Probationer sample and Lower M.C. sample: $\chi^2 = 0.65$; N.S.
Control sample and Lower M.C. sample: $\chi^2 = 1.12$; N.S.

groups gave more negative than positive responses. But there were no differences between the probationers and their matched controls, though the boys in the lower middle-class sample were significantly less enthusiastic than even the two other groups of boys. Of course, this finding tells us very little about the probationers' actual orientation to material goods and the possibilities of possessing them. In general, too, few respondents in any of the samples seemed particularly concerned about commercials, either positively or negatively.

The viewing situation

All respondents were asked whether they usually watched television in the same room as the one where they ate their meals. As Table 23 shows, probationers of both sexes were significantly more likely to do this than was the case with either of the two other samples, but there were no differences here between the control and lower middle-class samples.[2]

Table 23

Room where viewing takes place

View in room where meals eaten	*Probationer sample*	*Control sample*	*Lower M.C. sample*
	MALES		
Yes	140 (51%)	33 (34%)	45 (36%)
No	135 (49%)	64 (66%)	81 (64%)
Total	275 (100%)	97 (100%)	126 (100%)

Probationer sample and Control sample: $\chi^2 = 7.55$; $p < 0.01$
Probationer sample and Lower M.C. sample: $\chi^2 = 7.43$; $p < 0.01$
Control sample and Lower M.C. sample: $\chi^2 = 0.01$; N.S.

	FEMALES		
Yes	31 (61%)	18 (41%)	22 (39%)
No	20 (39%)	26 (59%)	34 (61%)
Total	51 (100%)	44 (100%)	56 (100%)

Control sample and Lower M.C. sample: $\chi^2 = 2.98$; N.S.
Probationer sample and Lower M.C. sample: $\chi^2 = 4.11$; $p < 0.05$
Probationer sample and Control sample: $\chi^2 = 0.00$; N.S.

Table 24

Freedom to view programmes

Allowed to watch any programme	*Probationer sample*	*Control sample*	*Lower M.C. sample*
	MALES		
Yes	164 (58%)	68 (69%)	63 (50%)
No	117 (42%)	30 (31%)	64 (50%)
Total	281 (100%)	98 (100%)	127 (100%)

Probationer sample and Control sample: $\chi^2 = 3.27$; N.S.
Probationer sample and Lower M.C. sample: $\chi^2 = 2.37$; N.S.
Control sample and Lower M.C. sample: $\chi^2 = 8.10$; $p < 0.01$

	FEMALES		
Yes	29 (55%)	29 (63%)	33 (57%)
No	24 (45%)	17 (37%)	25 (43%)
Total	53 (100%)	46 (100%)	58 (100%)

Probationer sample and Control sample: $\chi^2 = 0.40$; N.S.
Probationer sample and Lower M.C. sample: $\chi^2 = 0.00$; N.S.
Control sample and Lower M.C. sample: $\chi^2 = 0.19$; N.S.

All respondents were asked whether or not they were allowed to watch whatever programmes they wanted. Table 24 indicates that, so far as males were concerned, a lower proportion of probationers than of control-sample members said that they had this sort of freedom, though the difference was not statistically significant. A significantly smaller number of lower middle-class males were allowed to view anything they wanted to than was true of the control sample, and this is what one would have predicted on the basis of prior research findings: middle-class parents would usually be expected to exercise greater control over their children's media exposure than working-class parents. However, this was not the case with girls; indeed, here the probationer sample registered the smallest proportion with complete freedom to view.

Table 25

Person deciding what is viewed

Person who decides	*Probationer sample*	*Control sample*	*Lower M.C. sample*
	MALES		
Father	77 (28%)	40 (41%)	37 (29%)
Mother	63 (23%)	11 (11%)	23 (18%)
Parents	28 (10%)	5 (5%)	10 (8%)
Other	112 (40%)	41 (42%)	57 (45%)
Total	280 (100%)	97 (100%)	127 (100%)

Probationer sample and Control sample: $\chi^2 = 10.97$; $p < 0.05$
Probationer sample and Lower M.C. sample: $\chi^2 = 1.81$; N.S.
Control sample and Lower M.C. sample: $\chi^2 = 4.70$; N.S.

Person who decides	*Probationer sample*	*Control sample*	*Lower M.C. sample*
	FEMALES		
Father	13 (25%)	13 (30%)	14 (24%)
Mother	6 (12%)	7 (16%)	8 (14%)
Parents	3 (6%)	3 (7%)	5 (9%)
Other	30 (58%)	21 (48%)	31 (53%)
Total	52 (100%)	44 (100%)	58 (100%)

Probationer sample and Control sample: $\chi^2 = 0.98$; df $= 2$; N.S.
Probationer sample and Lower M.C. sample: $\chi^2 = 0.45$; df $= 2$; N.S.
Control sample and Lower M.C. sample: $\chi^2 = 0.59$; df $= 2$; N.S.

All respondents were asked who decided what was viewed in their house, and Table 25 shows that, in the case of boys, the control sample differed sharply from the probationer sample. In the former group the father was stated to be responsible for making decisions four times as often as the mother, whereas in the latter group the father was mentioned only slightly more often than the mother. The fact that the father seems in this context to have played a less important role for probationers than for control-sample members may in part be due to the fact that fewer probationers were living with their natural fathers (as seen above in Table 8), and to the fact that the father, when actually present, played a less dominant role within the household. A good deal of research, of course, has suggested that delinquents' fathers play a less central role in their households than is the case with other households in roughly the same social stratum. However, as Table 25 also shows, there was no significant difference between the probationer and lower middle-class samples. However, the lesser prominence of the father among the latter group of boys may well be due to quite other factors than those suggested in the case of the probationers. Even among the lower strata of the middle class, the control of children may be left much more to the mother, while the father's work may serve to segregate him more from the round of household activity. There were however no similar inter-sample differences in who made decisions about viewing in the girls' households.

Viewing in company

Respondents were explicitly asked during the interview whether they ever watched television 'in order to have some company'. Table 26 shows that lower middle-class boys were somewhat less likely than members of the two other samples to say that they did watch television for this reason, but this was not the case with girls. Although the percentage difference between the probationers and members of the control sample was not significant for either sex, there seems to be a tendency for probationers to have claimed that this reason applied to them less often than was the case with the control sample. Perhaps this trend can be explained in terms of the fact that the households to which probationers belonged were large (as shown above in Table 15), so that they were less likely to find occasion to use television as a replacement for actual human company.

Respondents were later on asked to say whether they watched television in the company of a range of other persons, 'Very often', 'Fairly often', 'Not so often' or 'Never'. Scores of 3, 2, 1 and 0 were assigned to these responses, and the mean scores for the different samples are shown in the following tables in this section.

Table 27 indicates that there were no differences for males in the frequency of viewing with fathers, though the girls in the probationer sample said that they viewed with their fathers less frequently than was the case in the control sample. There were no inter-sample differences at all in the case of viewing with mothers (Table 28), though there seemed to be a tendency for male probationers to state a preference for viewing with their mothers more commonly than was true of boys in either of the other samples (Table 29).

Table 26

Watching television in order to have some company

Probationer sample	*Control sample*		*Lower M.C. sample*
	MALES		
158 (56%)	66 (67%)		58 (46%)
Probationer sample and Control sample:		$Z = 1.93$; N.S.	
Probationer sample and Lower M.C. sample:		$Z = 1.98$; $p < 0.05$	
Control sample and Lower M.C. sample:		$Z = 3.25$; $p < 0.01$	
	FEMALES		
30 (57%)	27 (59%)		35 (60%)
Probationer sample and Control sample:		$Z = 0.21$; N.S.	
Probationer sample and Lower M.C. sample:		$Z = 0.40$; N.S.	
Control sample and Lower M.C. sample:		$Z = 0.17$; N.S.	

Table 27

Viewing in company of father

Probationer sample	*Control sample*		*Lower M.C. sample*
	MALES		
1.79	1.82		1.79
Probationer sample and Control sample:		$t = 0.22$; N.S.	
Probationer sample and Lower M.C. sample:		$t = 0.02$; N.S.	
Control sample and Lower M.C. sample:		$t = 0.23$; N.S.	
	FEMALES		
1.58	2.07		1.83
Probationer sample and Control sample:		$t = 2.18$; $p < 0.05$	
Probationer sample and Lower M.C. sample:		$t = 1.23$; N.S.	
Control sample and Lower M.C. sample:		$t = 1.27$; N.S.	

Table 28

Viewing in company of mother

Probationer sample	*Control sample*		*Lower M.C. sample*
	MALES		
2.10	1.94		2.15
Probationer sample and Control sample:		$t = 1.32$; N.S.	
Probationer sample and Lower M.C. sample:		$t = 0.52$; N.S.	
Control sample and Lower M.C. sample:		$t = 1.81$; N.S.	
	FEMALES		
2.09	2.22		2.03
Probationer sample and Control sample:		$t = 0.58$; N.S.	
Probationer sample and Lower M.C. sample:		$t = 0.35$; N.S.	
Control sample and Lower M.C. sample:		$t = 1.00$; N.S.	

Table 29

Preferences for viewing television with mother

Probationer sample	*Control sample*		*Lower M.C. sample*
	MALES		
49 (17%)	9 (9%)		9 (7%)
Probationer sample and Control sample:		$Z = 1.95$; N.S.	
Probationer sample and Lower M.C. sample:		$Z = 2.77$; $p < 0.01$	
Control sample and Lower M.C. sample:		$Z = 0.58$; N.S.	
	FEMALES		
8 (15%)	10 (22%)		5 (9%)
Probationer sample and Control sample:		$Z = 0.86$; N.S.	
Probationer sample and Lower M.C. sample:		$Z = 1.06$; N.S.	
Control sample and Lower M.C. sample:		$Z = 1.89$; N.S.	

In the light of the size of the probationers' households, it is not surprising that they should view more frequently in the company of siblings than was true of other groups. However, as Table 30 indicates, this only held for boys. Viewing with 'somebody else in the family' becomes less frequent as one goes from the probationer sample to the control sample to the middle-class sample. This holds for both sexes, as Table 31 shows. However, only the middle-class males differed significantly from other groups. The differences indicated by this table are no doubt due to the differing sizes of households already referred to, with larger households making for greater opportunities to view in company. Female probationers viewed with 'anybody else we haven't mentioned so far' significantly more often than girls in the two other samples, but there was no indication of such a difference for boys. These comparisons are shown in Table 32.

There were no inter-sample differences in the frequency with which members of different samples viewed with friends (Table 33). However, when data from the question on viewing with girlfriends or boyfriends are analysed, a more complicated picture emerges, as Table 34 shows. The average frequency for boys rises as one moves from probationers across to middle-class boys, with a difference between these two extreme groups which is highly significant statistically. Although none of the girls' inter-sample differences is significant, it is worth noting that the mean score for the female probationers is higher than that for the two other samples so that the mean between-sex difference is greatest within the probationer sample. The data suggest that male probationers were less involved with members of the opposite sex than were members of the two other samples. Conversely, female probationers were at least as competent at personal relationships with the opposite sex as members of other groups, if not more so. (The larger number of 17–20 year olds somewhat vitiates this comparison, however.) Other data, to be reported in a later section, support this pattern of sex-linked differences.

Table 30
Viewing in company of siblings

Probationer sample	*Control sample*		*Lower M.C. sample*
	MALES		
2.15	1.90		1.69
Probationer sample and Control sample:		$t = 1.97$; $p < 0.05$	
Probationer sample and Lower M.C. sample:		$t = 3.90$; $p < 0.001$	
Control sample and Lower M.C. sample:		$t = 1.40$; N.S.	
	FEMALES		
2.04	2.04		1.91
Probationer sample and Control sample:		$t = 0.03$; N.S.	
Probationer sample and Lower M.C. sample:		$t = 0.57$; N.S.	
Control sample and Lower M.C. sample:		$t = 0.59$; N.S.	

Table 31
Viewing in company of some other member of family

Probationer sample	*Control sample*		*Lower M.C. sample*
	MALES		
1.05	0.92		0.62
Probationer sample and Control sample:		$t = 0.99$; N.S.	
Probationer sample and Lower M.C. sample:		$t = 3.64$; $p < 0.001$	
Control sample and Lower M.C. sample:		$t = 2.22$; $p < 0.05$	
	FEMALES		
0.98	0.74		0.67
Probationer sample and Control sample:		$t = 1.11$; N.S.	
Probationer sample and Lower M.C. sample:		$t = 1.55$; N.S.	
Control sample and Lower M.C. sample:		$t = 0.37$; N.S.	

For each individual member of each sample, total 'viewing with' scores were calculated by summing across the range of persons referred to in the interview, and group means were then calculated from these summed scores. These means are shown in Table 35. Within each sample, girls appear to view television in company slightly more frequently than boys. Within sexes, probationers achieve the highest mean scores, with control-sample members on average viewing in company rather less frequently, and lower middle-class adolescents less frequently still. Only the difference between

Table 32

Viewing in company of some other person

Probationer sample	*Control sample*	*Lower M.C. sample*
	MALES	
0.43	0.50	0.40

Probationer sample and Control sample: $t = 0.68$; N.S.
Probationer sample and Lower M.C. sample: $t = 0.38$; N.S.
Control sample and Lower M.C. sample: $t = 0.96$; N.S.

	FEMALES	
0.77	0.39	0.38

Probationer sample and Control sample: $t = 2.06$; $p < 0.05$
Probationer sample and Lower M.C. sample: $t = 2.32$; $p < 0.05$
Control sample and Lower M.C. sample: $t = 0.09$; N.S.

Table 33

Viewing in company of friends

Probationer sample	*Control sample*	*Lower M.C. sample*
	MALES	
1.00	1.04	0.87

Probationer sample and Control sample: $t = 0.36$; N.S.
Probationer sample and Lower M.C. sample: $t = 1.53$; N.S.
Control sample and Lower M.C. sample: $t = 1.70$; N.S.

	FEMALES	
1.13	1.22	1.21

Probationer sample and Control sample: $t = 0.49$; N.S.
Probationer sample and Lower M.C. sample: $t = 0.47$; N.S.
Control sample and Lower M.C. sample: $t = 0.06$; N.S.

Table 34

Viewing in company of girlfriend/boyfriend

Probationer sample	*Control sample*	*Lower M.C. sample*
	MALES	
0.28	0.43	0.62

Probationer sample and Control sample: $t = 1.68$; N.S.
Probationer sample and Lower M.C. sample: $t = 4.04$; $p < 0.001$
Control sample and Lower M.C. sample: $t = 1.67$; N.S.

	FEMALES	
0.62	0.43	0.50

Probationer sample and Control sample: $t = 1.12$; N.S.
Probationer sample and Lower M.C. sample: $t = 0.78$; N.S.
Control sample and Lower M.C. sample: $t = 0.40$; N.S.

Table 35

Overall frequency of viewing with other persons

Probationer sample	*Control sample*		*Lower M.C. sample*
	MALES		
8.80	8.54		8.14
Probationer sample and Control sample:		$t = 0.78$; N.S.	
Probationer sample and Lower M.C. sample:		$\mathbf{t = 2.28;\ p < 0.01}$	
Control sample and Lower M.C. sample:		$t = 1.13$; N.S.	
	FEMALES		
9.23	9.11		8.53
Probationer sample and Control sample:		$t = 0.24$; N.S.	
Probationer sample and Lower M.C. sample:		$t = 1.47$; N.S.	
Control sample and Lower M.C. sample:		$t = 1.12$; N.S.	

boys in the probationer and middle-class samples reaches a level of statistical significance, however. It can be noted that the ranking of samples on this overall 'viewing with' variable parallels that for the mean sizes of households shown in Table 15. On the basis of a number of the preceding tables in this section it is clear that a large proportion of the variance shown in Table 35 can be accounted for in terms of the differential *availability* of persons with whom to view television. The data on boys' viewing with girlfriends, however, run strongly counter to the overall pattern.

Amount of viewing

In general, probationers did not differ from members of the control sample in the numbers of hours they spent viewing television. Lower middle-class respondents viewed in general for significantly fewer hours than was the case with the two other samples. This latter difference was predictable on the basis of earlier research, though there are some theoretical grounds upon which it might have been expected that the probationers would view more than control-sample members. These arguments are referred to later on.

Two sets of questions dealing with amount of viewing were employed during the interview. Although the two resulting estimates of the number of hours of television viewed per week differ quite considerably, the differences are in the same direction for each group of respondents, so that they both seem to be of use in *comparing* groups, despite the fact that it is difficult to place much trust in either measure as an estimate of the *actual*, rather than the *relative*, amount of viewing.

The first question consisted of a chart with three separate columns for weekdays, Saturdays, and Sundays. Each column was marked off in one-hour periods, and the person being interviewed was asked to say whether he or she was likely to view *some* television during each period, first for weekdays, then for Saturdays, and finally for Sundays. To provide a crude estimate of the number of hours of viewing per week, the number of hours during which at least *some* television was viewed on weekdays was multiplied by five, and the totals for Saturday and Sunday added to this figure. Calculated in this way, the mean number of hours for each group are shown in Table 36.

Table 36

Number of clock hours during which some television viewed

Day	*Probationer sample*	*Control sample*	*Lower M.C. sample*
		MALES	
Weekday	4.68	4.74	4.20
Probationer sample and Control sample: $t = 0.27$; N.S.			
Probationer sample and Lower M.C. sample: $t = 2.40$; $p < 0.05$			
Control sample and Lower M.C. sample: $t = 2.28$; $p < 0.05$			
Saturday	6.31	6.77	6.33
Probationer sample and Control sample: $t = 1.39$; N.S.			
Probationer sample and Lower M.C. sample: $t = 0.07$; N.S.			
Control sample and Lower M.C. sample: $t = 1.18$; N.S.			
Sunday	5.87	5.97	5.83
Probationer sample and Control sample: $t = 0.34$; N.S.			
Probationer sample and Lower M.C. sample: $t = 0.15$; N.S.			
Control sample and Lower M.C. sample: $t = 0.36$; N.S.			
Total per week	35.57	36.46	32.74
Probationer sample and Control sample: $t = 0.59$; N.S.			
Probationer sample and Lower M.C. sample: $t = 2.08$; $p < 0.05$			
Control sample and Lower M.C. sample: $t = 3.25$; $p < 0.01$			
		FEMALES	
Weekday	4.34	4.24	4.09
Probationer sample and Control sample: $t = 0.27$; N.S.			
Probationer sample and Lower M.C. sample: $t = 0.83$; N.S.			
Control sample and Lower M.C. sample: $t = 0.46$; N.S.			
Saturday	4.75	4.89	4.90
Probationer sample and Control sample: $t = 0.23$; N.S.			
Probationer sample and Lower M.C. sample: $t = 0.28$; N.S.			
Control sample and Lower M.C. sample: $t = 0.02$; N.S.			
Sunday	5.32	5.26	5.36
Probationer sample and Control sample: $t = 0.11$; N.S.			
Probationer sample and Lower M.C. sample: $t = 0.08$; N.S.			
Control sample and Lower M.C. sample: $t = 0.21$; N.S.			
Total per week	31.77	31.35	30.69
Probationer sample and Control sample: $t = 0.16$; N.S.			
Probationer sample and Lower M.C. sample: $t = 0.50$; N.S.			
Control sample and Lower M.C. sample: $t = 0.31$; N.S.			

CHART 1

Effects of sex and day of the week
on number of clock hours during which some television viewed
(data points from Table 36)

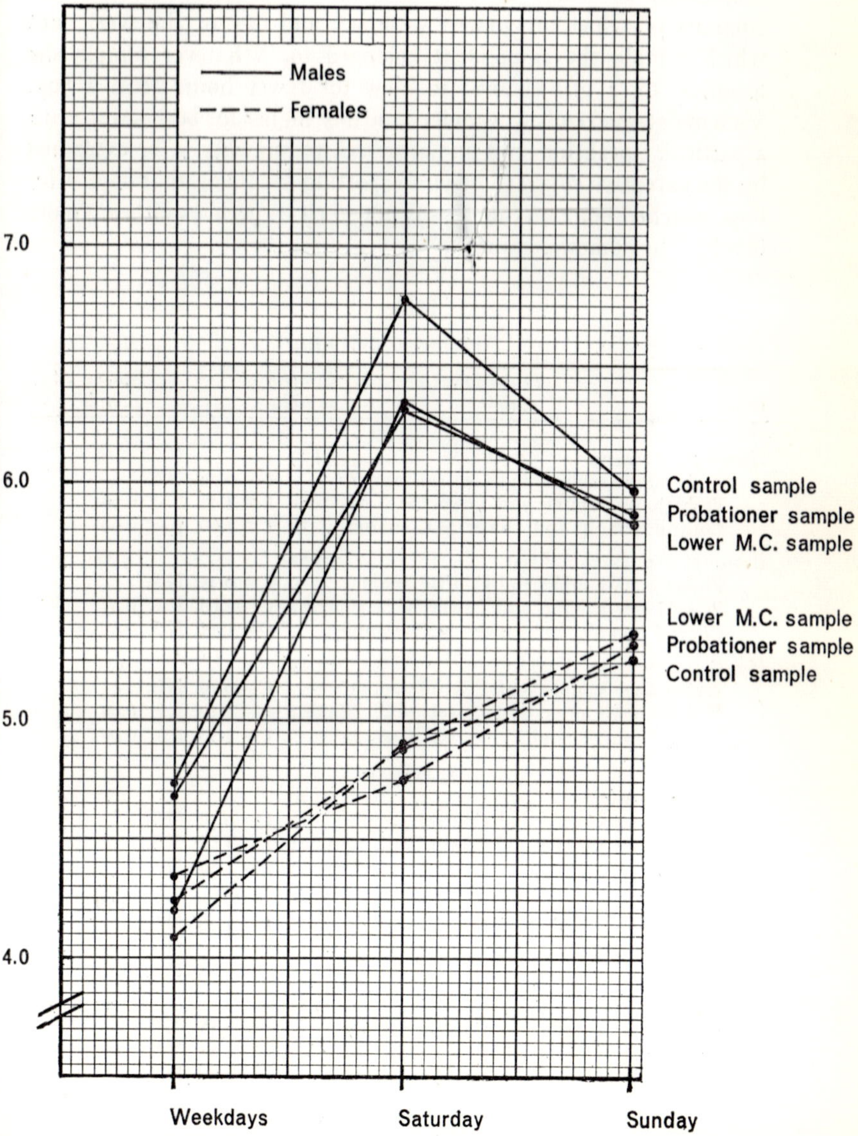

For males there are no significant differences between the probationer sample and the control sample, while on weekdays and for the total week the middle-class boys come out with totals significantly lower than those characteristic of either of the two other samples, although for Saturdays and Sundays all three means were effectively the same. So far as girls are concerned, there are no significant inter-sample differences at all. The overall situation is brought out more graphically in Chart 1. The variables which really affected the amounts of television viewed were sex and day of the week, not which sample the respondent belonged to. Whatever sample she belonged to, a girl claimed to view for fewer hours than a boy. Viewing at weekends was understandably higher for both sexes, with a particular peak for boys on Saturdays, presumably to be explained by the extensive coverage given to sporting events on that day. Why boys watched more on Sundays is not so clear; perhaps girls are more involved in domestic chores.

Table 37

Amount of viewing

Index	*Probationer sample*	*Control sample*	*Lower M.C. sample*
MALES			
Hours per evening	3.91	4.05	3.39
Probationer sample and Control sample:			$t = 0.79$; N.S.
Probationer sample and Lower M.C. sample:			$t = 3.45$; $p < 0.001$
Control sample and Lower M.C. sample:			$t = 3.66$; $p < 0.001$
Evenings per week	5.91	5.82	5.52
Probationer sample and Control sample:			$t = 0.45$; N.S.
Probationer sample and Lower M.C. sample:			$t = 2.09$; $p < 0.05$
Control sample and Lower M.C. sample:			$t = 1.35$; N.S.
Hours per week	23.83	24.23	19.31
Probationer sample and Control sample:			$t = 0.29$; N.S.
Probationer sample and Lower M.C. sample:			$t = 3.76$; $p < 0.001$
Control sample and Lower M.C. sample:			$t = 3.50$; $p < 0.001$
FEMALES			
Hours per evening	4.08	4.20	3.67
Probationer sample and Control sample:			$t = 0.36$; N.S.
Probationer sample and Lower M.C. sample:			$t = 1.54$; N.S.
Control sample and Lower M.C. sample:			$t = 1.89$; N.S.
Evenings per week	6.02	5.43	5.69
Probationer sample and Control sample:			$t = 1.76$; N.S.
Probationer sample and Lower M.C. sample:			$t = 1.01$; N.S.
Control sample and Lower M.C. sample:			$t = 0.68$; N.S.
Hours per week	24.64	23.72	21.84
Probationer sample and Control sample:			$t = 0.38$; N.S.
Probationer sample and Lower M.C. sample:			$t = 1.33$; N.S.
Control sample and Lower M.C. sample:			$t = 0.81$; N.S.

CHART 2

Effects of sex and sample
on self-rated assessments of exposure to television
(data points from Table 37)

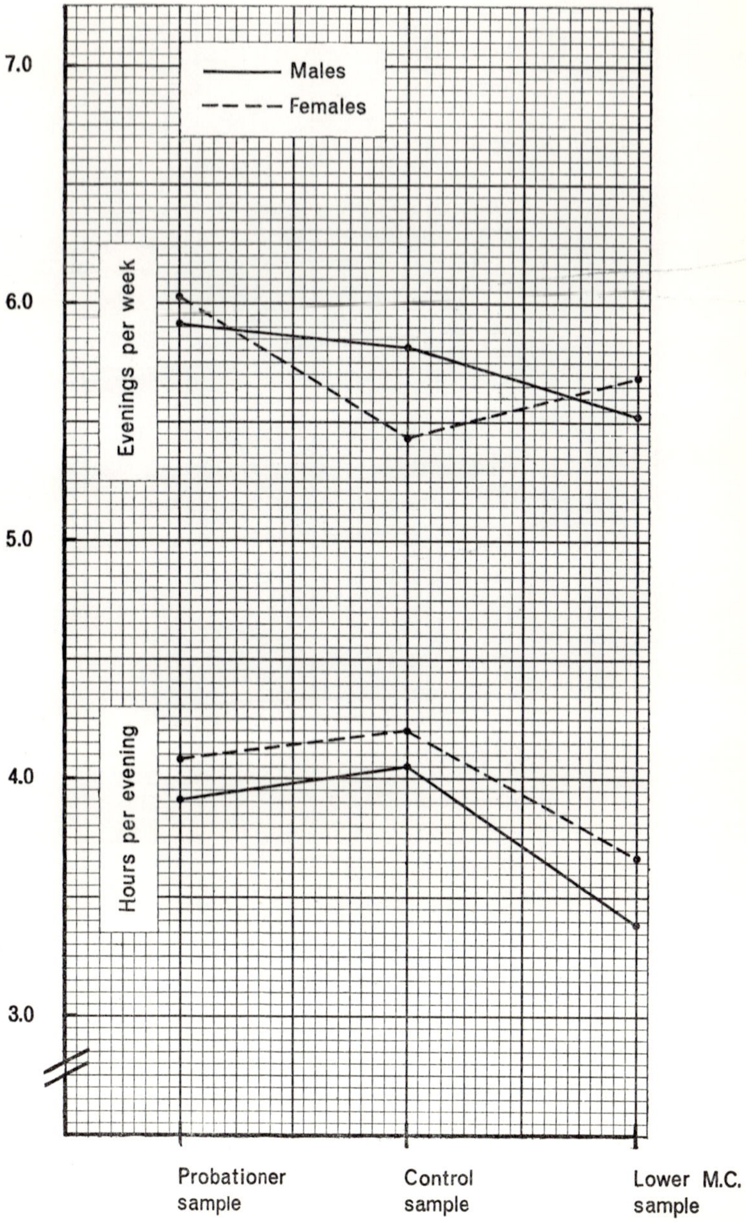

CHART 3

Effects of sex and sample
on two viewing indices
(data points from Tables 36 and 37)

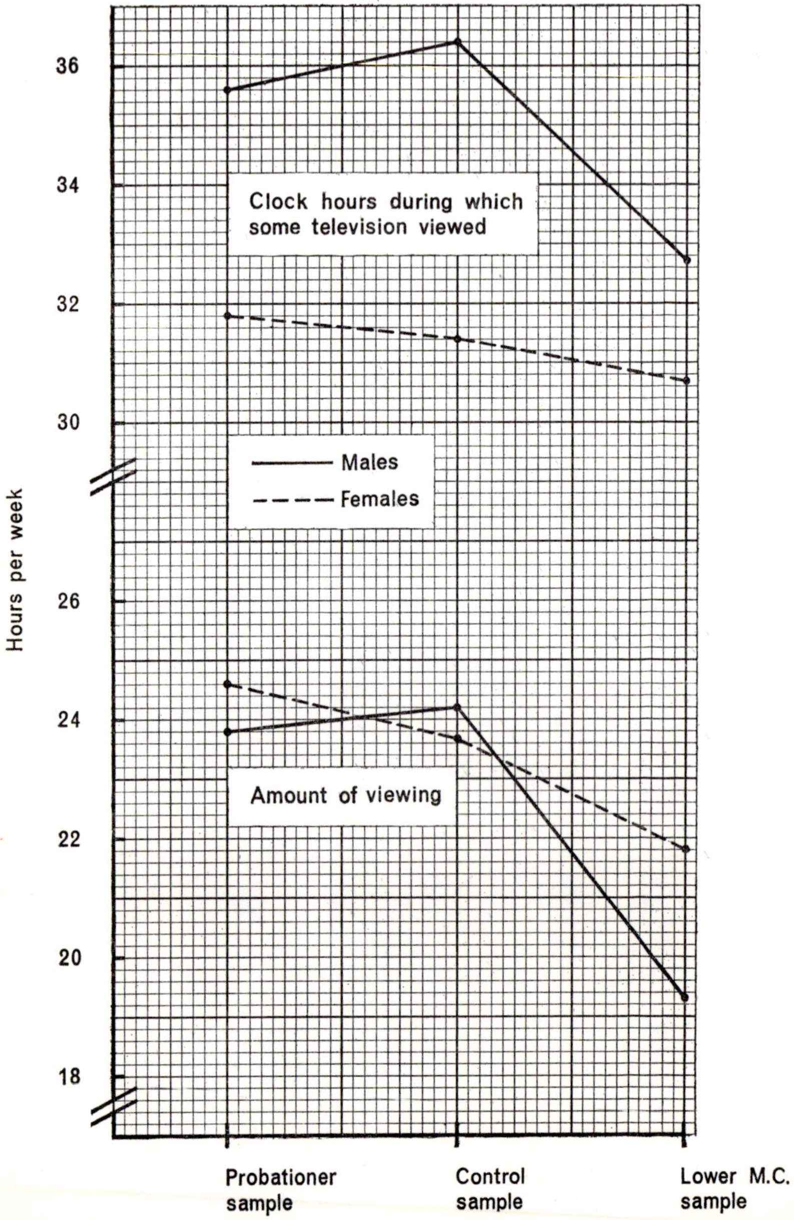

To provide a second estimate of the amount of viewing during the week, respondents were asked to estimate how many hours they would be likely to view during an average evening, and on how many evenings during the week they would watch television. The number of hours viewed per average evening was then multiplied by the number of viewing evenings per average week to provide an estimate of how many hours an individual respondent might view during the entire week.

Table 37 shows the resulting averages for each group of respondents. For boys, once again, there are no significant differences between the probationer sample and the control sample, while lower middle-class boys claimed to view for significantly fewer hours per evening than members of these two other samples. Of the three samples, the middle-class boys also viewed television for the fewest number of evenings on average, although only the comparison with the probationer sample is statistically significant. The highly significant differences between the middle-class boys and the members of the other groups in terms of overall amount of viewing per week naturally reflects the differences in terms of evenings and hours per evening. As with the earlier questions on viewing, there were no significant differences between the three samples of girls.

In Chart 2, it can be seen that the pattern of inter-sample differences in terms of the average number of viewing hours per evening is similar for both sexes, while in terms of the average number of evenings per week girls in the control sample appear to represent a deviant group.

Chart 3 brings together data on the two different estimates of the total amount of viewing during the week. For both sexes, the pattern of inter-sample differences is the same for both types of estimate, though the rise from probationer sample to control sample for boys is not seen in the lines for girls. The averages derived from the three-column chart, and shown at the top in the diagram, also separate the sexes more distinctly than do the scores arrived at by multiplying hours by evenings. The fact that the estimates derived from the three-column chart are so much higher than those given by the multiplication of numbers of hours and evenings is no doubt mainly accounted for by the fact that a respondent's indication that he was likely to watch *some* television during a particular clock hour on a particular day of the week was treated in all cases as equal to a full hour's viewing. Thus the 'Hours per week' entries in Table 37 are probably the best available estimates of actual television exposure. Since the two estimates of total television exposure per week were essentially measuring different things, it is not surprising that the linear correlations between them are low: 0.38 for the probationer sample, 0.52 for the control sample, and 0.49 for the lower middle-class sample. However, all these correlations are statistically highly significant.

Up-to-date information on the amount of time children and adolescents spend viewing television is hard to come by. However, the 'Hours per week' entries in Table 37 do not seem too far out of line with data published by Granada Television.[3] On the basis of fieldwork conducted in the early months of 1960, a figure of approximately 20 hours of viewing per week was arrived at for children in the 5–15 age range (considerably younger on average than the samples studied here). This is roughly comparable to the figures of from 19 to 25 hours suggested by Table 37. The Granada figure for BBC-only homes was somewhat lower (about 16 hours). However, respondents in the present survey were not asked which channels the television sets they viewed at home could receive.

Finding that members of the lower middle-class sample viewed for fewer hours than other respondents causes no surprise. A number of studies have established class differences in television viewing behaviour of this sort, while Geiger and Sokol have attempted to demonstrate that watching television is less congruent with middle-class norms than with working-class norms.[4] However, it is interesting that differences should emerge so clearly in the present study between the working-class control samples and the *lower* middle-class sample. Of course, the academically better qualified members of this latter group may have received more homework than did members of the two other samples or spent more time on it, and thus just had fewer opportunities to view.

The fact that the probationers and the control sample did not differ in amount of viewing also merits attention. A number of earlier empirical studies suggest reasons why the probationers might be expected to have watched more than members of the control sample.

Coleman's report on the relationship between amount of mass media exposure and success (or rather the lack of it) within the competitive High School system in the United States suggests that the media are used more heavily by those who have failed to win recognition among peers.[5] A similar picture emerges from Riley and Riley's study of the media behaviour of younger American adolescents.[6] The Rileys found that adolescents who were not members of peer groups were heavier users of the media than those who were successful members of peer groups. If the probationers perceived themselves as failures in terms of some system of values which was of importance to them, then one might expect that they too would have indulged in heavier media usage, just like the American adolescents in the two studies referred to. It seems reasonable to assume, despite data presented above in Tables 6 and 7, that the probationers may indeed have been relative failures in terms of the value system of the school. Thus to the extent that they recognized academic and school values, the probationers might have been expected to react to their lack of success in terms of heavier use of television. Again, male probationers' apparent lack of success in their dealings with girlfriends, suggested by Table 34, might be taken as indirect evidence of failure in terms of the adolescent value system, and hence might have been expected to result in more hours spent viewing television. The fact that these theoretical expectations are *not* borne out by the actual data calls in question the appropriateness in this context of the sort of theory which links failure in terms of recognized goals with heavier media usage.

We shall suggest later on that the delinquents did not recognize the middle-class values of the school, and from their own point of view are probably not to be regarded as failures. Again, we shall suggest that the male probationers ought not to be looked on as failures where girlfriends are concerned, but must probably be seen as members of all-male groups deliberately avoiding feminine company. There are findings to support both of these contentions, and these findings can in each case be linked to general theories about the male delinquent's values. We shall thus be led to conclude that the Riley-Coleman theory of compensatory viewing is inappropriate here.

At the moment we can merely say that the probationers did not view any more or any less than other working-class adolescents. But having examined the available data on gross amounts of television exposure, it is now possible to turn and look more closely at the respondents' orientation to the actual content of television programmes.

Recently, a good deal of attention has been paid by mass communications researchers to the uses to which different categories of persons put the various media and their content. This way of looking at mass media audiences is commonly known as the 'uses and gratifications' or 'functional' approach. It is distinguished from analysis of the direct effects of communication by its posing the question "What do people do with mass media content?" rather than the question "What does mass media content do to people?"[7] A body of findings from the present study lends itself rather readily to analysis using such a 'uses and gratifications' approach, and a fairly consistent pattern of inter-sample differences emerges. The next two chapters deal mainly with two particularly significant uses of television content.

IV

EXCITEMENT AND MASCULINITY

Excitement and relaxation

During the interviews, each respondent was invited to name up to six *sorts* of programmes which he or she particularly liked. A few actual titles of programmes were given in response to this question as well as some purely evaluative adjectives (for example, 'good'). Although the majority of respondents found some difficulty in giving a full list of six programme types, a fair amount of useful data was collected. On average, the girls in each sample gave a longer list of favourite types than the boys: male probationers gave 3.64 answers each on average, and female probationers 3.83 answers; in the control sample the figures for boys and girls were 3.50 and 3.91 respectively; while the lower middle-class boys provided 3.77 answers each on average, and the girls 3.95. Individual answers to the question were initially categorized into 40 classes. The use of 40 coding categories led to relatively low frequencies in most of them. Thus analysis of these data inevitably involved grouping together into somewhat broader classes what could be reasonably considered as homogeneous sets of categories.

A set of thirteen types of programme seemed to have in common the fact that each of the types could easily be thought of as offering a measure of immediate emotional stimulation to the viewer. The labels given to the thirteen categories involved were: 'Science fiction', 'Mystery', 'Police', 'Crime', 'Westerns', 'Spy', 'Thrillers', 'Danger', 'Aggression', 'Murder', 'War films', 'Exciting' and 'Frightening'. A reference to *any* of these types of programmes was labelled as a reference to an 'Exciting' programme.

Table 38 shows that there are good grounds for saying that, so far as boys are concerned, probationers had a stronger preference for 'Exciting' programmes than did boys in the matched control sample. This latter group also showed a stronger preference for such material than boys in the lower middle-class sample. In fact, all the three possible differences are statistically significant. It is worth noticing too that girls in each of the three samples exhibit a lesser liking for programmes of this kind than do the boys, and that the inter-sex difference seems particularly strong among probationers. This is of course also reflected in the fact that the girls in the control sample do not differ significantly here from the probationers, although the lower middle-class girls show a significantly lower preference for 'Exciting' programmes than girls in the control sample. The girls' pattern here almost exactly repeats that found for boys.

Table 38

*Preferences for 'Exciting' programmes**

Probationer sample	*Control sample*	*Lower M.C. sample*
	MALES	
404 (42%)	121 (35%)	135 (28%)
(N = 971)	(N = 343)	(N = 479)

Probationer sample and Control sample: $Z = 2.06$; $p < 0.05$
Control sample and Lower M.C. sample: $Z = 4.97$; $p < 0.001$
Probationer sample and Lower M.C. sample: $Z = 2.17$; $p < 0.05$

Probationer sample	*Control sample*	*Lower M.C. sample*
	FEMALES	
65 (32%)	61 (34%)	57 (25%)
(N = 203)	(N = 180)	(N = 229)

Probationer sample and Control sample: $Z = 0.39$; N.S.
Probationer sample and Lower M.C. sample: $Z = 1.64$; N.S.
Control sample and Lower M.C. sample: $Z = 2.00$; $p < 0.05$

* Many respondents gave more than one relevant answer; up to six answers could be given to the question. The same proviso applies to subsequent tables.

After they had said what sorts of programmes they particularly liked, respondents were asked what sorts of programmes they particularly *disliked*. In the case of this question, up to four sorts of programmes were recorded. Exactly the same set of categories was used to code the answers as was used for the question about preferred types of programmes. Table 39 shows the frequency with which dislike for the broad class of 'Exciting' programmes was expressed. None of the inter-sample comparisons approaches statistical significance, although the girls in each sample appeared to dislike this sort of programme more than the boys, a finding which fits in well with the fact that girls also appeared to *like* this sort of programme less. Ignoring inter-sample differences, the broad pattern of sex differences was to be expected, of course.

One way of summing up the general attitude of each group of respondents to 'Exciting' programmes is to calculate the ratio of the number of expressed preferences for this kind of content to the number of expressed dislikes. When this is done, male probationers

Table 39
Dislikes of 'Exciting' programmes

Probationer sample	*Control sample*		*Lower M.C. sample*
	MALES		
45 (8%)	15 (8%)		21 (7%)
(N = 574)	(N = 187)		(N = 289)
Probationer sample and Control sample:		Z = 0.08; N.S.	
Probationer sample and Lower M.C. sample:		Z = 0.30; N.S.	
Control sample and Lower M.C. sample:		Z = 0.30; N.S.	
	FEMALES		
14 (11%)	12 (13%)		15 (12%)
(N = 125)	(N = 94)		(N = 126)
Probationer sample and Control sample:		Z = 0.35; N.S.	
Probationer sample and Lower M.C. sample:		Z = 0.17; N.S.	
Control sample and Lower M.C. sample:		Z = 0.19; N.S.	

clearly show the greatest degree of liking, with a ratio of 8.98. Boys in the control sample show a ratio of 8.07, and those in the lower middle-class sample a ratio of 6.43. Each group of girls has a lower ratio than the corresponding group of boys: 4.64 for probationers, a somewhat higher figure of 5.08 for the controls, and the noticeably low figure of 3.80 for the lower middle-class girls. There seems little doubt, in other words, that a liking for what have been termed 'Exciting' programmes is connected with being a boy.

At one point during the interview, each respondent was asked to give a single word to describe the sort of television programmes he or she wishes were shown more often, that is, his or her ideal television fare. In fact, some interviewees had considerable difficulty with this and similar questions, and many gave answers of more than one word. However, the answers could be categorized, and three of the categories finally employed were labelled 'Exciting', 'Thrilling' and 'Adventurous'. It seemed reasonable to group these three types of answer together and treat the occurrence of any one of them as an expressed desire that television should provide excitement. Table 40 shows that lower middle-class boys expressed this sort of wish somewhat less often than members of either of the two other samples, but that the situation for girls was apparently quite different. However, the numbers of responses on which the girls' percentages are based are extremely low. None of the six differences reaches the level of statistical significance.

Parallel to the question asking respondents to state a word to describe television as it ideally would be for them, an earlier question had asked for a word to describe what feeling or emotion respondents would ideally like television to produce *in* them. Here again, 'Excited' labelled one of the coding categories used. Table 41 indicates that once again the boys in the probationer sample gave answers of this sort more often than either of the two other groups of boys, and that there is a linear trend from the probationers to the lower middle-class boys. However, the difference between even these two extreme groups barely reaches a level of statistical significance. It is somewhat surprising that the lower middle-class girls registered the highest rate of response of all six groups, though of course one has no way of knowing from the data what particular types of television content respondents had in mind: boys and girls are likely to have been referring to rather different sorts of output.

Before being asked to say how they would ideally like television to make them feel, respondents were asked to give an indication of how television actually made them feel. They were requested to provide one word to complete the sentence "When I watch television it makes me feel . . . " One quite frequent response was 'Relaxed' or some similar word, and although it occurred less frequently, this type of answer was also sometimes given to the question which immediately followed in the interviews. This concerned the feeling that respondents would *ideally* like television to produce in them. Tables 42 and 43 show that the frequency with which this type of response was given rose for each question as one moved from probationers to lower middle-class boys, with the latter group standing by itself. Once again, the pattern for the girls is far less clear. However, the differing frequencies with which 'Relaxed' was given as a response to each of the two questions about the feelings produced by watching television indicate that there was some felt discrepancy between television as it is and television as it could ideally be imagined.

If we combine the frequencies shown in Tables 42 and 43, and express the figure for actual television as a ratio of the figure for ideal television, then we shall have an index characterizing the extent to which respondents would have liked television to be a *non*-relaxing stimulus. The calculations give boys in the probationer sample the highest ratio of 6.33, the control sample boys a somewhat lower figure of 5.25, and the lower middle-class boys a ratio of only 2.44. The corresponding index figures for girls are 2.71, 1.33 and 5.67.

So far then as boys are concerned, a wish to use television as a means of relaxation is weakest among the probationers and strongest among members of the middle-class sample. Conversely, as the tables presented earlier in this chapter indicate, probationers looked to television for excitement more than did the lower middle-class boys. Boys in the control sample fell somewhere in the middle in terms of both excitement and relaxation. Male probationers seem to have manifested a distinctive pattern of orientations to television. Why was this?

The theoretical literature on delinquency does not furnish a fully adequate account of excitement-seeking amongst delinquents. However, the fact that the probationers studied here had a dispropor-

Table 40

Spontaneous mentions of 'Exciting', 'Thrilling', or 'Adventurous' as description of ideal television fare

Probationer sample	*Control sample*	*Lower M.C. sample*
	MALES	
67 (28%)	19 (27%)	22 (20%)
(N = 239)	(N = 70)	(N = 112)
Probationer sample and Control sample:	Z = 0.15; N.S.	
Probationer sample and Lower M.C. sample:	Z = 1.68; N.S.	
Control sample and Lower M.C. sample:	Z = 1.18; N.S.	
	FEMALES	
12 (28%)	12 (43%)	17 (38%)
(N = 43)	(N = 28)	(N = 45)
Control sample and Lower M.C. sample:	Z = 1.30; N.S.	
Probationer sample and Lower M.C. sample:	Z = 0.99; N.S.	
Control sample and Lower M.C. sample:	Z = 0.43; N.S.	

Table 41

Spontaneous mentions of 'Excitement' as the ideal feeling produced by television

Probationer sample	*Control sample*	*Lower M.C. sample*
	MALES	
28 (12%)	7 (8%)	6 (5%)
(N = 235)	(N = 84)	(N = 114)
Probationer sample and Control sample:	Z = 0.90; N.S.	
Probationer sample and Lower M.C. sample:	Z = 1.97; $p < 0.05$	
Control sample and Lower M.C. sample:	Z = 0.86; N.S.	
	FEMALES	
4 (10%)	3 (7%)	7 (14%)
(N = 41)	(N = 42)	(N = 49)
Probationer sample and Control sample:	Z = 0.43; N.S.	
Probationer sample and Lower M.C. sample:	Z = 0.65; N.S.	
Control sample and Lower M.C. sample:	Z = 1.09; N.S.	

Table 42

Spontaneous mentions of 'Relaxation' as the feeling produced by television

Probationer sample	*Control sample*		*Lower M.C. sample*
	MALES		
57 (23%)	21 (24%)		44 (38%)
(N = 246)	(N = 87)		(N = 115)
Probationer sample and Control sample:		$Z = 0.18$; N.S.	
Probationer sample and Lower M.C. sample:		$Z = 2.98$; $p < 0.01$	
Control sample and Lower M.C. sample:		$Z = 2.13$; $p < 0.05$	
	FEMALES		
19 (38%)	12 (29%)		17 (34%)
(N = 50)	(N = 42)		(N = 50)
Probationer sample and Control sample:		$Z = 0.95$; N.S.	
Probationer sample and Lower M.C. sample:		$Z = 0.42$; N.S.	
Control sample and Lower M.C. sample:		$Z = 0.56$; N.S.	

Table 43

Spontaneous mentions of 'Relaxation' as the ideal feeling produced by television

Probationer sample	*Control sample*		*Lower M.C. sample*
	MALES		
9 (4%)	4 (5%)		18 (16%)
(N = 235)	(N = 84)		(N = 114)
Probationer sample and Control sample:		$Z = 0.37$; N.S.	
Probationer sample and Lower M.C. sample:		$Z = 3.92$; $p < 0.001$	
Control sample and Lower M.C. sample:		$Z = 2.44$; $p < 0.05$	
	FEMALES		
7 (17%)	9 (21%)		3 (6%)
(N = 41)	(N = 42)		(N = 49)
Probationer sample and Control sample:		$Z = 0.50$; N.S.	
Probationer sample and Lower M.C. sample:		$Z = 1.65$; N.S.	
Control sample and Lower M.C. sample:		$Z = 2.15$; $p < 0.05$	

tionate liking for excitement in the area of television does nevertheless square with other reported findings about juvenile delinquents.

Mays, in his study of the social background of delinquency in Liverpool, notes that boys who told him of the delinquencies they had committed in the past often also said that it was the attendant feelings of danger, thrill and adventure which made their exploits so attractive.[1] Mays notes too that the Gluecks, in their massive research into the concomitants of delinquency in the United States, "have already drawn attention to the fact that delinquents are above the average in their zest for exciting recreation and dangerous play."[2] It certainly seems possible that the probationers' heightened liking for excitement in television programming reflects similar characteristics. Mays himself draws the parallel between delinquent behaviour and various types of fantasy play when he writes that: "We know that most children love the excitement of battle by watching the games they play. Such games as cowboys and Indians, roundheads and royalists, smugglers and excisemen, gunmen and police, involve conflict and fear in fantasy and minister to the same psychological needs as climbing walls or steep cliffs or delinquent breaking into locked premises."[3] Of course, this does not explain either why only some boys express this sort of need in delinquent activity, or why delinquents have a heightened need for excitement.

Miller sees the search for excitement as a characteristic feature of lower-class cultures, and suggests that the committing of exciting delinquent acts should be set against long periods of boredom and inaction.[4] However, in the present context Miller's account may explain the relatively low importance of excitement for the lower middle-class boys better than it explains the slighter differences between the delinquents and their controls.

Heroes and aggressiveness

A good deal of research interest focuses on the types of television characters and personalities whom different categories of viewer admire, choose as models for emulation, or identify with. A number of questions were asked during each interview in order to produce data in this area, and some of the resulting findings are presented at this point.

It seems likely that the probationers' apparent search for excitement from television was closely linked with their orientation to the fictional characters seen in some programmes. They probably found programmes exciting to the extent that there was a hero figure in the story with whose exciting exploits they were able to identify. The data certainly suggest that television heroes emphasizing masculine behaviour patterns had particular salience for male delinquents.

All respondents were asked to list the titles of up to six actual television programmes of which they were particularly fond. When the list was complete, or the respondent was unable to think of any more favourites, the interviewer went back to the beginning of the list of programme titles and asked the *reason* for each one being especially liked. One identifiable category of response referred to the hero or leading character in the programme in terms of admiration for his skill, prowess or (occasionally) moral qualities. Table 44 shows that boys in the probationer sample were significantly more likely to give this type of answer than other boys, while there were no differences between the girls' groups.

The responses to the question asked for reasons justifying the choice of favourite programmes were also coded in terms of the particular *aspect* of the programme referred to, and Table 45 shows that in this respect also male probationers were significantly more likely to refer to a hero figure than other boys. There were no significant differences between the girls' groups, although girls appear to have made this sort of response more often than boys.

Some further weight is added to the suggestion that heroes in television drama programmes had particular salience for male probationers by the findings presented in Table 46. All respondents were asked to name or describe the two scenes on television which they particularly remembered, or which had made a particular impression on them. The table shows that male probationers were more likely to refer to an identifiable hero figure than were other boys (although only the difference between the probationer and control samples is statistically significant). There was a similar difference between male probationers and control-sample boys in the frequency with which answers falling into the category labelled 'Feats of prowess' occurred, and answers of this sort often referred implicitly to a programme hero figure. Here, however, lower middle-class boys gave more responses of this sort than either of the other boys' groups.

Table 44

Admiration of hero as reason for liking favourite programmes

Probationer sample	*Control sample*	*Lower M.C. sample*
	MALES	
103 (11%)	10 (3%)	15 (3%)
(N = 901)	(N = 371)	(N = 563)

Probationer sample and Control sample: $Z = 4.98$; $p < 0.001$
Probationer sample and Lower M.C. sample: $Z = 6.00$; $p < 0.001$
Control sample and Lower M.C. sample: $Z = 0.03$; N.S.

	FEMALES	
9 (5%)	3 (2%)	9 (4%)
(N = 197)	(N = 176)	(N = 250)

Probationer sample and Control sample: $Z = 1.57$; N.S.
Probationer sample and Lower M.C. sample: $Z = 0.52$; N.S.
Control sample and Lower M.C. sample: $Z = 1.17$; N.S.

Table 45

References to heroes as reason for liking favourite programmes

Probationer sample	*Control sample*		*Lower M.C. sample*
	MALES		
76 (10%) (N = 757)	16 (5%) (N = 294)		18 (4%) (N = 407)
Probationer sample and Control sample:		Z = 2.37; p < 0.05	
Probationer sample and Lower M.C. sample:		Z = 3.36; p < 0.001	
Control sample and Lower M.C. sample:		Z = 0.62; N.S.	
	FEMALES		
16 (10%) (N = 156)	24 (15%) (N = 160)		23 (12%) (N = 188)
Probationer sample and Control sample:		Z = 1.27; N.S.	
Probationer sample and Lower M.C. sample:		Z = 0.58; N.S.	
Control sample and Lower M.C. sample:		Z = 0.75; N.S.	

Table 46

*Memories of scenes on television featuring heroes (males)**

Memory	*Probationer sample*	*Control sample*	*Lower M.C. sample*
Hero or personality	23 (10%) (N = 221)	2 (2%) (N = 91)	6 (5%) (N = 126)
Probationer sample and Control sample:		Z = 2.43; p < 0.05	
Probationer sample and Lower M.C. sample:		Z = 1.83; N.S.	
Control sample and Lower M.C. sample:		Z = 0.99; N.S.	
Feats of prowess	24 (11%) (N = 221)	2 (2%) (N = 91)	24 (19%) (N = 126)
Probationer sample and Control sample:		Z = 2.52; p < 0.05	
Probationer sample and Lower M.C. sample:		Z = 2.13; p < 0.05	
Control sample and Lower M.C. sample:		Z = 3.77; p < 0.001	

* Answer frequencies for females too low for analysis.

The fact that male probationers remembered and liked the type of hero who appears in television programmes appears to agree with findings reported by Bailyn.[5] Working with 10- and 11-year-old children in the Boston area in the United States, Bailyn found that television material (and other mass media material) featuring specifically *aggressive* heroes was liked particularly by boys (though not by girls) who had numerous difficulties in their relationships with others and who manifested what was termed 'rebellious independence'. These latter characteristics, and others which Bailyn identified, have of course also been shown to be associated with juvenile delinquency. Indeed, Bailyn noted that the factors which predicted a liking for material featuring aggressive heroes were similar to those isolated by the Gluecks in their study of the social-psychological concomitants of delinquency.[6]

From the data collected in the present study, it is impossible to say whether the heroes and other similar media figures whom the male probationers particularly liked were specifically aggressive in behaviour, or whether they were perceived as such. However, the titles of programmes mentioned by respondents when asked to name their favourite television fare, and common knowledge of the British, 'mid-Atlantic' and American series material that was being transmitted in large quantities during the period when the present study was being conducted, both make this a reasonable assumption.

One finding adds a little weight to this position. As already indicated, three of the categories used to classify the sorts of television programmes mentioned as favourites were labelled 'Aggression', 'Murder', and 'War films'. These three categories were later subsumed into a broader class re-labelled 'Aggressive' programmes. As Table 47 indicates, girls in general liked such programmes less than boys, while amongst boys such programmes became less popular as one moved from the probationer to the lower middle-class sample. The difference between the probationers and their controls is not statistically significant, but it is in the direction one would have predicted on the basis of the other findings outlined above.

The nature of the present study does not allow us to draw any firm conclusions about the functions or consequences of the male

Table 47

Preferences for 'Aggressive' programmes

Probationer sample	*Control sample*	*Lower M.C. sample*
	MALES	
129 (13%)	36 (10%)	13 (3%)
(N = 971)	(N = 343)	(N = 479)

Probationer sample and Control sample: $Z = 1.34$; N.S.
Probationer sample and Lower M.C. sample: $Z = 6.37$; $p < 0.001$
Control sample and Lower M.C. sample: $Z = 4.66$; $p < 0.001$

Probationer sample	*Control sample*	*Lower M.C. sample*
	FEMALES	
17 (8%)	9 (5%)	9 (4%)
(N = 203)	(N = 180)	(N = 229)

Probationer sample and Control sample: $Z = 1.31$; N.S.
Probationer sample and Lower M.C. sample: $Z = 1.94$; N.S.
Control sample and Lower M.C. sample: $Z = 0.52$; N.S.

probationers' liking for television hero figures. It is possible that probationers liked these figures because they provided a means whereby aggression generated in socially frustrating situations could be vicariously discharged. It is equally possible that viewing material featuring these sorts of characters resulted in tension levels being raised rather than lowered. Indeed, the differing degrees to which members of the three samples claimed to look for excitement in their television viewing may suggest that boys in the probationer sample were likely to be more highly stimulated than boys in the two other samples. However, we have no data by which to test such an hypothesis. Further, it is only possible to speculate about whether such presumed increases in tension during viewing were likely to last beyond the actual screening of the programme.

The probationers' greater liking for what Bailyn labelled 'aggressive hero material' may well underlie both their more frequent references to programme heroes and their stronger preference for what were labelled 'Exciting' programmes earlier in this book. If so, the same kind of explanation may serve to account for both sets of differences.

Several sociological theories of delinquency suggest that deviant behaviour on the part of male adolescents is the result of an over-emphasis on the male role. Parsons traces back such an over-emphasis to the sex-role relationships supposedly characteristic of the restricted, nuclear family in industrial societies.[7] Since the father's work is removed from the domestic scene, the male child inevitably models itself on the mother in the early years. Later, a stage is reached at which the child, realizing that such an identification is inappropriate, rejects the female role entirely and tends toward what Parsons terms 'compulsive masculinity'. The extreme result of such over-compensation may be an over-aggressive behaviour pattern. In some cases, this may lead to acts classifiable as deviant or delinquent, although socially approved forms of competitive behaviour in the work sphere may in other cases serve equally well as an outlet. Parsons does not himself suggest why it is that the reaction-formation process he sketches should result in socially stigmatized behaviour in only some cases, and not universally. However, in families where the father has been absent (as in many of the probationers' households in the present study), or has failed entirely to provide an adequate male identification figure, the reaction against female identification may be sharper, and the search for alternative male identification figures more intense. The fact that television hero figures appear to have had greater salience for male probationers may be explicable in terms of some such theory.

Parsons' theory is, of course, a very general one, since family circumstances as he describes them are characteristic of most social classes. In one sense it is not a very powerful theory, since one wants to know why it is that working- or lower-class male adolescents are particularly prone to commit delinquent acts manifesting some degree of masculine aggressiveness. A number of writers have advanced theories which may be seen as extensions of Parsons' argument.

Miller, for example, lists 'Toughness' as one of the 'focal concerns' of lower-class culture, and characterizes the appropriate behaviour in terms of exaggerated masculinity. The theoretical account of this feature of lower-class culture is best given in Miller's own words:

> "The genesis of the intense concern over 'toughness' in lower-class culture is probably related to the fact that a significant proportion of lower-class males are reared in a predominantly female household, and lack a consistently present male figure with whom to identify and from whom to learn essential components of a 'male' role. Since women serve as a primary object of identification during pre-adolescent years, the almost obsessive lower-class concern with 'masculinity' probably resembles a type of compulsive reaction-formation."[8]

The appropriateness of such a theory in the case of our own delinquent sample is immediately apparent. The probationers studied were often not living with their natural fathers, and even where this was the case the father was frequently unemployed or in irregular employment, and in ill-health. Many of the probationers probably saw their fathers as providing inadequate identification figures. In its turn, this may account for the particular salience which television heroes appeared to have had for the probationers. However, here as at other points where we seek to account for the delinquents' viewing patterns, this suggestion must be read as an hypothesis generated from the data, and not as a fully proven statement of fact.

V

INFORMATION AND EDUCATION

Television and informal learning

Some television programmes are specifically designed to provide information or instruction. Even programmes whose main function is supposedly to entertain may be sources of incidental learning about a wide range of topics. It is therefore quite reasonable to think of television as a medium which lends itself to being used as a source of knowledge. Further, more intelligent adolescents will be particularly likely to use television in this way. Not only will the more intelligent be better equipped to understand and enjoy specifically informative programmes: through the status which intellectual ability gives within the formal educational system, the pursuit of knowledge will be likely to develop into a self-rewarding activity, not only within the classroom, but in relation to the mass media as well.

There are in fact a number of audience studies which show that intelligence is related to the use of the media as sources of knowledge. This relationship can be illustrated with data collected in the present study. In Table 48, all respondents are placed in one of two categories according to whether their intelligence quotients fall below or above the average for the entire sample. The table shows that those in the higher intelligence category were significantly more likely to make some reference to gaining knowledge when asked to say, without prompting, *why* they watched television; and the higher intelligence group were also significantly more likely to refer to knowledge or information when describing the *sort* of television they would ideally like to view. In the case of both indices of the use of television as a source of knowledge, the differences between intelligence categories are considerable, and highly significant statistically.

It will be recalled that the boys in the probationer sample did not differ from members of the control sample in terms of average intelligence scores. Consequently, basing predictions purely on intelligence, there are strong grounds for hypothesizing that no differences should emerge between the probationers and the control sample in terms of their use of television as a source of knowledge and information. Indeed, since the probationers' average intelligence was marginally higher than that of the control sample members (significantly so for

Table 48

Level of intelligence and the use of television as a source of knowledge

		Level of intelligence Low	High
Gaining knowledge spontaneously mentioned as reason for watching television	+	47 (19%)	90 (35%)
	–	200 (81%)	168 (65%)
		247 (100%)	258 (100%)
		$\chi^2 = 15.25$; $p < 0.001$	
Reference to information in description of ideal television programme	+	7 (3%)	33 (13%)
	–	240 (97%)	225 (87%)
		247 (100%)	258 (100%)
		$\chi^2 = 15.82$; $p < 0.001$	

girls), there might even be grounds for predicting that the probationers would be slightly heavier users of television as a source of knowledge. In fact, the evidence points in the other direction.

All the respondents in the survey were specifically asked whether they watched television in order to learn something. As Table 49 shows, so far as boys are concerned the probationer sample stands by itself in terms of the percentages giving a positive response to this question, with the lowest figure of the three. In the case of girls, the differences between the probationers and the control sample, and between the control sample and the lower middle-class sample, are both significant. With both sexes, the proportion of respondents saying they watched television in order to learn something rises progressively as one moves from the probationer sample to the control sample and then to the lower middle-class sample. Female members of the probationer sample seem to score particularly low, and female members of the middle-class sample particularly high. More important, the differences between the probationers and their matched controls run counter to what one would predict on the basis of respondents' intelligence scores.

So far as boys are concerned, just the same pattern of differences emerges when an analysis is made of answers to an open-ended question about reasons for watching television (Table 50). Again, only the difference between the probationers and members of the lower middle-class sample is significant, but the linear trend is clear enough. Lower middle-class girls score almost as high as lower middle-class boys, but female probationers do not differ significantly from girls in the control sample.

Table 49

Watching television in order to learn something

Probationer sample	*Control sample*		*Lower M.C. sample*
	MALES		
229 (81%)	87 (89%)		114 (90%)
Probationer sample and Control sample:		$Z = 1.67$; N.S.	
Probationer sample and Lower M.C. sample:		$Z = 2.12$; $p < 0.05$	
Control sample and Lower M.C. sample:		$Z = 0.24$; N.S.	
	FEMALES		
33 (62%)	39 (85%)		56 (97%)
Probationer sample and Control sample:		$Z = 2.51$; $p < 0.05$	
Probationer sample and Lower M.C. sample:		$Z = 4.53$; $p < 0.001$	
Control sample and Lower M.C. sample:		$Z = 2.12$; $p < 0.05$	

Table 50

Spontaneous mentions of 'Knowledge' or 'Interest' as a reason for watching television

Probationer sample	*Control sample*		*Lower M.C. sample*
	MALES		
118 (18%)	50 (23%)		83 (28%)
(N = 639)	(N = 222)		(N = 301)
Probationer sample and Control sample:		$Z = 1.31$; N.S.	
Probationer sample and Lower M.C. sample:		$Z = 3.40$; $p < 0.001$	
Control sample and Lower M.C. sample:		$Z = 1.48$; N.S.	
	FEMALES		
21 (18%)	18 (17%)		39 (27%)
(N = 119)	(N = 103)		(N = 145)
Probationer sample and Control sample:		$Z = 0.03$; N.S.	
Probationer sample and Lower M.C. sample:		$Z = 1.78$; N.S.	
Control sample and Lower M.C. sample:		$Z = 1.74$; N.S.	

Of the 40 categories used to code answers to the previously discussed questions on the *sorts* of programmes respondents particularly liked and disliked, seven categories could be placed together to form a broader class labelled 'Educational and cultural'. The categories subsumed into this broader class were labelled: 'News', 'Political', 'Talking', 'Religious', 'Educational', 'Hobbies' and 'Classical music'. Although not all the sorts of programmes alluded to by these category headings are specifically concerned with the dissemination of formal knowledge, there are good grounds for supposing that preferences for them will be correlated with levels of intelligence. In terms of the present study, this again means that there should be no differences between members of the probationer and control samples. In fact, as with the other questions indexing the use of television as a source of knowledge, male probationers expressed preferences for 'Educational and cultural' programmes significantly less often than was the case with the control-sample boys. Table 51 shows that this difference was statistically significant, as was that between control-sample boys and

Table 51

Preferences for 'Educational and cultural' programmes

Probationer sample	*Control sample*	*Lower M.C. sample*
	MALES	
77 (8%)	46 (13%)	89 (19%)
(N = 971)	(N = 343)	(N = 479)
Probationer sample and Control sample:		$Z = 3.00$; $p < 0.01$
Probationer sample and Lower M.C. sample:		$Z = 5.99$; $p < 0.001$
Control sample and Lower M.C. sample:		$Z = 1.97$; $p < 0.05$
	FEMALES	
18 (9%)	9 (5%)	31 (14%)
(N = 203)	(N = 180)	(N = 229)
Probationer sample and Control sample:		$Z = 1.48$; N.S.
Probationer sample and Lower M.C. sample:		$Z = 1.53$; N.S.
Control sample and Lower M.C. sample:		$Z = 2.89$; $p < 0.01$

Table 52

Dislikes of 'Educational and cultural' programmes

Probationer sample	*Control sample*	*Lower M.C. sample*
	MALES	
184 (32%)	66 (35%)	106 (37%)
(N = 574)	(N = 187)	(N = 289)
Probationer sample and Control sample:		$Z = 0.82$; N.S.
Probationer sample and Lower M.C. sample:		$Z = 1.36$; N.S.
Control sample and Lower M.C. sample:		$Z = 0.31$; N.S.
	FEMALES	
60 (48%)	36 (38%)	45 (36%)
(N = 125)	(N = 94)	(N = 126)
Probationer sample and Control sample:		$Z = 1.43$; N.S.
Probationer sample and Lower M.C. sample:		$Z = 1.97$; $p < 0.05$
Control sample and Lower M.C. sample:		$Z = 0.39$; N.S.

boys in the lower middle-class sample. In fact, the linear trend seen in Tables 49 and 50 appears here once again. With girls, too, a familiar pattern re-emerges. Members of the middle-class sample prefer 'Educational and cultural' programmes significantly more than girls in either of the two other samples, while again girls in the probationer sample score slightly, though not significantly, higher than control-sample girls. This may well be due to their significantly higher mean intelligence.

'Educational and cultural' programmes were also mentioned by a number of respondents when they were asked to state the sorts of programmes which they *disliked*. Table 52 shows that here there were no differences between boys in the three samples, but that probationer-sample girls expressed significantly more dislikes for this sort of programming than was the case with the lower middle-class girls.

We can sum up a particular group's net attitude towards 'Educational and cultural' programmes by expressing their number of stated preferences as a ratio of their number of stated dislikes. A higher ratio will then express a greater liking for this sort of television output. The actual figures suggest that, among boys, members of the probationer sample were least enthusiastic, with a ratio of 0.42, and members of the lower middle-class sample most enthusiastic, with a ratio of 0.84. Control-sample boys fell between the two, with a ratio of 0.70. The girls' ratios were in each case lower than those of the corresponding groups of boys, with figures of 0.30 (probationers), 0.25 (control sample) and 0.69 (lower middle-class sample).

It will be recalled that respondents were invited to list by name up to six favourite programmes, and subsequently asked to explain why they liked the ones they mentioned. Two of the coding categories set up for the reasons given were labelled 'Interesting' and 'Informative'. The combined frequencies of responses falling into these two categories are shown in Table 53. There is again a rising trend from probationers to lower middle-class sample members among the boys, and with the girls too the middle-class respondents gave the relevant sorts of answer most commonly. The only difference which is statistically significant for boys is that between the two extreme groups.

Respondents gave the names of up to four disliked programmes, and were again asked to explain their choices. Two of the coding categories used to classify the reasons given were labelled 'Uninteresting' and 'Too much talking'. The combined frequencies with which replies of this kind were offered are shown in Table 54. Here the boys' frequencies *fall* as one moves from the probationers to the third sample, and the difference between the probationers and controls is statistically significant. However, the control-sample girls gave a significantly higher proportion of responses falling into one or other of the two relevant categories than was true of the female probationers.

Table 53

'Interesting' or 'Informative' as a reason for liking particular programmes

Probationer sample	*Control sample*	*Lower M.C. sample*
	MALES	
160 (18%)	76 (20%)	139 (25%)
(N = 901)	(N = 371)	(N = 563)

Probationer sample and Control sample: $Z = 1.14$; N.S.
Probationer sample and Lower M.C. sample: $Z = 3.19$; $p < 0.01$
Control sample and Lower M.C. sample: $Z = 1.50$; N.S.

Probationer sample	*Control sample*	*Lower M.C. sample*
	FEMALES	
31 (16%)	25 (14%)	61 (24%)
(N = 197)	(N = 176)	(N = 250)

Probationer sample and Control sample: $Z = 0.41$; N.S.
Probationer sample and Lower M.C. sample: $Z = 2.25$; $p < 0.05$
Control sample and Lower M.C. sample: $Z = 2.58$; $p < 0.01$

Table 54

'Uninteresting' or 'Too much talking' as a reason for disliking particular programmes

Probationer sample	*Control sample*	*Lower M.C. sample*
	MALES	
260 (38%)	68 (29%)	89 (25%)
(N = 689)	(N = 233)	(N = 352)

Probationer sample and Control sample: $Z = 2.36$; $p < 0.05$
Probationer sample and Lower M.C. sample: $Z = 4.03$; $p < 0.001$
Control sample and Lower M.C. sample: $Z = 1.04$; N.S.

Probationer sample	*Control sample*	*Lower M.C. sample*
	FEMALES	
38 (25%)	47 (39%)	47 (29%)
(N = 153)	(N = 119)	(N = 164)

Probationer sample and Control sample: $Z = 2.59$; $p < 0.01$
Probationer sample and Lower M.C. sample: $Z = 0.77$; N.S.
Control sample and Lower M.C. sample: $Z = 1.91$; N.S.

We have been assuming of course that replies which were coded as 'Interesting' or 'Informative', and 'Uninteresting' or 'Too much talking', had reference respectively to the two kinds of programmes previously called educational and cultural. However, we feel the assumption is a reasonable one. Here again, the ratio of a group's frequency of use of the first two kinds of term to their frequency of use of the latter two kinds of term can give us an indication of the group's net liking for programmes perceived to have a high knowledge content. When the ratios are calculated, the index among boys is lowest for the probationer sample (0.62), next lowest for the control sample (1.12) and highest for the lower middle-class group (1.56). The corresponding girls' ratios are 1.23, 1.88 and 0.77.

Two sorts of answer to the question asking respondents to describe the sort of television they wished were on the screen more often were again labelled 'Interesting' and 'Informative'. Here again, the boys' responses follow a familiar pattern, with the percentage of answers falling into one or other of these two categories rising as one moves from the probationer to the middle-class sample, while the percentage for each of the girls' samples is more nearly the same (Table 55).

It will be recalled that respondents were asked to say how they would ideally like television to make them feel. 'Interested' and 'Informed' were again the labels assigned to two of the categories into which answers could be coded, and it later seemed reasonable to group these together so as to provide somewhat larger numbers. Table 56 gives the frequencies with which 'Interested' or 'Informed' were

Table 55

Spontaneous mentions of 'Interesting' or 'Informative' as the description of ideal television fare

Probationer sample	*Control sample*	*Lower M.C. sample*
	MALES	
22 (9%)	9 (13%)	36 (32%)
(N = 239)	(N = 70)	(N = 112)

Probationer sample and Control sample: $Z = 0.90$; N.S.
Probationer sample and Lower M.C. sample: $Z = 5.40$; $p < 0.001$
Control sample and Lower M.C. sample: $Z = 2.93$; $p < 0.01$

Probationer sample	*Control sample*	*Lower M.C. sample*
	FEMALES	
10 (23%)	5 (18%)	9 (20%)
(N = 43)	(N = 28)	(N = 45)

Probationer sample and Control sample: $Z = 0.54$; N.S.
Probationer sample and Lower M.C. sample: $Z = 0.37$; N.S.
Control sample and Lower M.C. sample: $Z = 0.23$; N.S.

Table 56

Spontaneous mentions of 'Interested' or 'Informed' as the ideal feeling produced by television

Probationer sample	*Control sample*		*Lower M.C. sample*
	MALES		
28 (12%)	11 (13%)		26 (23%)
(N = 235)	(N = 84)		(N = 114)
Probationer sample and Control sample:		$Z = 0.28$; N.S.	
Probationer sample and Lower M.C. sample:		$Z = 2.64$; $p < 0.01$	
Control sample and Lower M.C. sample:		$Z = 1.74$; N.S.	
	FEMALES		
6 (15%)	7 (17%)		13 (27%)
(N = 41)	(N = 42)		(N = 49)
Probationer sample and Control sample:		$Z = 0.25$; N.S.	
Probationer sample and Lower M.C. sample:		$Z = 1.38$; N.S.	
Control sample and Lower M.C. sample:		$Z = 1.13$; N.S.	

used to characterize respondents' preferred emotional reactions to television. Here again, for boys, the pattern is clear enough: probationers mentioned 'Interested' or 'Informed' less often than boys in the other groups. For girls, the frequency of mention of 'Interested' or 'Informed' also rises as one moves from the probationer to the middle-class sample. However, all the differences are small, and none is statistically significant.

Educational television

The major focus of the questionnaire dealing with television behaviour was respondents' exposure to, and tastes for, the run of television programmes viewed at home. However, room was found towards the end of the schedule for two questions dealing with specifically instructional television seen at school, within the framework of the formal educational system. Table 57 indicates that significantly fewer male probationers claimed to have seen educational television programmes at school than was the case with the control sample. This was the only significant difference in exposure to this brand of television.[1]

Having established whether or not the interviewee had ever seen ETV programming, the interviewer then moved on to ask those who had had this experience to give their evaluations of it. The range of replies was fairly stereotyped, and it was possible to categorize most responses as either positive or negative, as shown in Table 58. The percentage of boys who had seen ETV programming and who

Table 57

Familiarity with educational television at school

Had watched television at school	*Probationer sample*	*Control sample*	*Lower M.C. sample*
	MALES		
Yes	161 (57%)	74 (76%)	81 (64%)
No	120 (43%)	24 (25%)	46 (36%)
Total	281 (100%)	98 (100%)	127 (100%)

Probationer sample and Control sample: $\chi^2 = 9.47$; $p < 0.01$
Probationer sample and Lower M.C. sample: $\chi^2 = 1.26$; N.S.
Control sample and Lower M.C. sample: $\chi^2 = 2.73$; N.S.

Had watched television at school	*Probationer sample*	*Control sample*	*Lower M.C. sample*
	FEMALES		
Yes	28 (53%)	19 (41%)	35 (60%)
No	25 (47%)	27 (59%)	23 (40%)
Total	53 (100%)	46 (100%)	58 (100%)

Probationer sample and Control sample: $\chi^2 = 0.89$; N.S.
Probationer sample and Lower M.C. sample: $\chi^2 = 0.37$; N.S.
Control sample and Lower M.C. sample: $\chi^2 = 3.00$; N.S.

evaluated it favourably rises as one moves from the probationers to the middle-class respondents, though only the difference between these two extreme groups approaches the level of statistical significance. Fewer female probationers, too, gave positive responses than was the case with the girls in the control sample, though the percentage of positive responses from the middle-class girls is lower than might be expected from other parts of the table. In the case of the boys, anyway, it seems that the attitudes which are reflected in earlier tables emerge here again in connection with the educational programmes seen at school. In fact the boys' reactions to schools' television can provide the starting point for an explanation of the rather consistent patterns of findings reported in this section.

The idea that a process of 'reaction formation' may be connected with juvenile delinquency has received considerable attention. The original formulation was put forward by Cohen, who argued that the working-class adolescent is torn between two paths.[2] He may cherish the sorts of aspirations suggested to him by the middle-class norms which dominate society at large, or he may reject both aspirations and norms. If an adolescent follows the first of these paths, then there is no conflict between his values and those of the larger, dominant social environment. If on the other hand a working-class adolescent for some reason rejects the values of middle-class society

Table 58

Evaluations of schools' television

Reaction to ETV	*Probationer sample*	*Control sample*	*Lower M.C. sample*
	MALES		
Positive	137 (66%)	62 (71%)	74 (80%)
Negative	51 (25%)	20 (23%)	15 (16%)
Other	19 (9%)	5 (6%)	4 (4%)
Total	207 (100%)	87 (100%)	93 (100%)

Probationer sample and Control sample: $\chi^2 = 1.19$; N.S.
Probationer sample and Lower M.C. sample: $\chi^2 = 5.74$; N.S.
Control sample and Lower M.C. sample: $\chi^2 = 1.26$; df = 1; N.S.

Reaction to ETV	*Probationer sample*	*Control sample*	*Lower M.C. sample*
	FEMALES		
Positive	23 (64%)	26 (76%)	32 (73%)
Negative	9 (25%)	7 (21%)	7 (16%)
Other	4 (11%)	1 (3%)	5 (11%)
Total	36 (100%)	34 (100%)	44 (100%)

Probationer sample and Control sample: $\chi^2 = 0.79$; df = 1; N.S.
Probationer sample and Lower M.C. sample: $\chi^2 = 0.37$; df = 1; N.S.
Control sample and Lower M.C. sample: $\chi^2 = 0.01$; df = 1; N.S.

(and additional theories are needed to account for the fact that this rejection is not automatic), then he is likely to experience a considerable degree of strain and anxiety. Although the adolescent has rejected middle-class values, these values still remain the dominant values of the society and of most of the social institutions he will encounter. In addition, the adolescent may himself have to some extent internalized these values, particularly in the school situation. Thus it is not enough merely to have rejected such values at a particular period: the process of rejection has to continue, and the constant pull towards 'giving in' and accepting the dominant social norms leads to an *over*-rejection of these same norms.[3] To keep the attractions of the dominant values at bay, the adolescent over-reacts against them. In Cohen's formulation, one possible manifestation of this over-action is delinquent behaviour.

The theory of reaction formation may serve to account for the present finding that the male probationers mentioned educational and cultural programmes among their favourite sorts less often than did members of the control sample, and disliked this kind of programming more. It has been pointed out often enough that the values of the school are essentially middle class. It would be natural to expect the adolescent who rejects middle-class values to reject equally strongly the content of formal education and related cultural materials. Among such materials, of course, may be television programmes perceived as educational.

It will be remembered that the male probationers were not significantly different from the control-sample boys in intelligence (Table 5), and of course this would lead one to predict that, other things being equal, the school performance of probationers should be much the same as that of boys in the control sample. And of course the data presented in Table 7 suggested that this was the case: judged by the norms of the school, probationers did at least as well as their intellectual ability would have led one to expect. These are findings which in themselves suggest that the probationers accepted the schools' values.

The data reported above, which indicated that probationers did not watch television for any more hours a week than was true of control-sample boys, leads us to the hypothesis that the male delinquents probably did not perceive themselves as 'failures' within the meritocratic school system. Rather, they appear to have detached themselves from the demands which it places on other children and adolescents. The fact that there were no differences in the *amounts* of television viewing characteristic of probationers and controls suggested that compensatory theories of the relationship between underachievement and media usage are inappropriate in the present context. By contrast the fact that the two samples were similar in mean intelligence, yet differed rather consistently in their attitudes to educational content on television, strongly suggests that a reaction-formation hypothesis is more tenable. At the least, it can be argued that the male probationers did not perceive themselves as failures or under-achievers, but rather rejected the norms by which others might have sought to categorize them as such.

People who are interested in obtaining and using knowledge are also likely to engage in a good deal of discursive thinking. Indeed, we cannot really claim to know anything until we are able to deal analytically with what we know. Conversely, those who are not very interested in knowledge for its own sake will have little time for discursive thinking. And of course the two things interact, so that those who are not very concerned with knowledge lose their ability to think analytically. In the next chapter we present a set of findings which suggest that the probationers probably analysed what they said about television less than did their matched controls, and were also less skilful at thinking discursively about what they had seen.

VI

CONVERSATION AND COGNITIVE POVERTY

Conversation about television

The analysis of the survey data revealed a number of differences between samples in the extent to which respondents talked about television with other persons. In many cases, probationers claimed to talk about television less often than did members of the control sample. There were also several differences in this respect between the control sample and the middle-class sample, but the pattern here was more complex. Some of the differences are fairly easily accounted for in terms of the different types of households characteristic of the three samples, but there are also findings which cannot be explained in these terms.

Respondents were asked how often they talked to a range of different persons, and invited in each case to choose one of four possible answers: 'Very often', 'Fairly often', 'Not so often' or 'Never'. Scores of 3, 2, 1, and 0 were assigned to these answers, and the means for groups calculated. It is these means which are presented in the tables in this section.

Table 59 shows that male probationers talked to their fathers about television significantly less frequently than was true for boys in the control sample. Remembering that significantly fewer probationers were living with their fathers (as shown above in Table 8), this difference is perfectly understandable. Middle-class boys did not differ significantly from those in the control sample. So far as girls are concerned, here the lower middle-class sample stands by itself, with the highest index of frequency of conversation; but none of the girls' inter-sample differences reaches the level of statistical significance.

So far as talking about television to mothers is concerned, Table 60 shows that the picture is very similar to that for fathers, with boys in the probationer sample claiming to talk significantly less often than boys in either of the two other samples. Since a significantly smaller proportion of male probationers were actually living with their mothers, this situation is again what was to be expected. While in each of the three samples the average frequency of conversation with fathers was lower for girls than for boys, in the case of talking to mothers the situation was reversed, with girls talking to mothers

Table 59

Frequency of conversation about television with father

Probationer sample	*Control sample*		*Lower M.C. sample*
	MALES		
0.89	1.21		1.13
Probationer sample and Control sample:		$t = 3.25$; $p < 0.01$	
Probationer sample and Lower M.C. sample:		$t = 2.65$; $p < 0.01$	
Control sample and Lower M.C. sample:		$t = 0.76$; N.S.	
	FEMALES		
0.79	0.83		1.12
Probationer sample and Control sample:		$t = 0.18$; N.S.	
Probationer sample and Lower M.C. sample:		$t = 1.76$; N.S.	
Control sample and Lower M.C. sample:		$t = 1.58$; N.S.	

Table 60

Frequency of conversation about television with mother

Probationer sample	*Control sample*		*Lower M.C. sample*
	MALES		
0.96	1.27		1.25
Probationer sample and Control sample:		$t = 2.94$; $p < 0.01$	
Probationer sample and Lower M.C. sample:		$t = 3.71$; $p < 0.001$	
Control sample and Lower M.C. sample:		$t = 0.13$; N.S.	
	FEMALES		
1.26	1.35		1.38
Probationer sample and Control sample:		$t = 0.49$; N.S.	
Probationer sample and Lower M.C. sample:		$t = 0.79$; N.S.	
Control sample and Lower M.C. sample:		$t = 0.19$; N.S.	

more than boys. It is perhaps worth suggesting that this situation fits in well enough with one pattern of household interaction which has been suggested as characteristic of working-class subcultures.

The finding that probationers talked to their parents about television less than was true of members of the control sample bears out data reported by Andry.[1] In a study in which he compared 80 male delinquents with 80 carefully matched controls, Andry found that the delinquent boys experienced "less adequate environmental communication with their parents" and "less adequate psychological communication."[2] It thus seems possible that the probationers in the present study would have been found to talk less to their parents about *other* topics besides television. However, the fact that mass media content is not discussed with parents may have particular consequences. We speculate about these consequences at the end of this section.

While the fact that probationers talked about television less with their parents than did members of the control sample is readily enough accounted for in terms of various types of parental absence, the inter-sample comparisons brought out in Table 61 are clearly not amenable to this sort of interpretation. Although neither of the differences is significant, probationers of both sexes talked less often to siblings about television than was true of the control sample; yet, as shown earlier in Table 15, probationers had a significantly larger number of siblings than was the case for members of the control sample. Thus it could be argued that a larger number of siblings would mean a greater number of opportunities for talking to them about television, and hence that frequency of conversation about television with siblings would indeed be higher in the probationer sample. It certainly seems possible to account for the differences between the control sample and the middle-class sample by an argument of this nature, since members of the latter group again had a significantly smaller number of brothers and sisters than was true of the non-delinquent working-class sample.

Table 61

Frequency of conversation about television with siblings

Probationer sample	*Control sample*	*Lower M.C. sample*
	MALES	
1.24	1.47	1.10

Probationer sample and Control sample: $t = 1.82$; N.S.
Probationer sample and Lower M.C. sample: $t = 1.17$; N.S.
Control sample and Lower M.C. sample: $t = 2.57$; $p < 0.05$

Probationer sample	*Control sample*	*Lower M.C. sample*
	FEMALES	
1.26	1.50	1.43

Probationer sample and Control sample: $t = 1.08$; N.S.
Probationer sample and Lower M.C. sample: $t = 0.86$; N.S.
Control sample and Lower M.C. sample: $t = 0.32$; N.S.

The possibility that the mere *availability* of people to talk to about television did not actually lead to probationers making use of such opportunities is again suggested by Table 62. Male probationers talked to 'somebody else in the family' significantly less often than did members of the matched control sample, and the same pattern appears for girls, although the difference is not statistically significant; yet, as shown above in Table 15, the households to which probationers belonged were significantly larger than those characteristic of the control sample. Just the same comment might be made about the findings reported in Table 63: on the basis of household size, it would have been expected that probationers would talk about television more often to 'anybody we haven't mentioned so far' than control-sample members. Yet the data exhibit a trend which is just the opposite to this.

Table 62

Frequency of conversation about television with some other member of family

Probationer sample	*Control sample*	*Lower M.C. sample*
	MALES	
0.42	0.72	0.59

Probationer sample and Control sample: $t = 3.16$; $p < 0.01$
Probationer sample and Lower M.C. sample: $t = 2.02$; $p < 0.05$
Control sample and Lower M.C. sample: $t = 1.13$; N.S.

Probationer sample	*Control sample*	*Lower M.C. sample*
	FEMALES	
0.43	0.63	0.66

Probationer sample and Control sample: $t = 1.12$; N.S.
Probationer sample and Lower M.C. sample: $t = 1.45$; N.S.
Control sample and Lower M.C. sample: $t = 0.14$; N.S.

Table 63

Frequency of conversation about television with some other person

Probationer sample	*Control sample*	*Lower M.C. sample*
	MALES	
0.27	0.39	0.44

Probationer sample and Control sample: $t = 1.39$; N.S.
Probationer sample and Lower M.C. sample: $t = 2.23$; $p < 0.05$
Control sample and Lower M.C. sample: $t = 0.47$; N.S.

Probationer sample	*Control sample*	*Lower M.C. sample*
	FEMALES	
0.49	0.50	0.41

Probationer sample and Control sample: $t = 0.06$; N.S.
Probationer sample and Lower M.C. sample: $t = 0.48$; N.S.
Control sample and Lower M.C. sample: $t = 0.53$; N.S.

The argument that the mere lack of available persons to talk to is not a sufficient explanation of why probationers conversed less about television is strengthened considerably when data on the probationers' relationships with their peers is examined.

Early on in the interview, each respondent was taken through a list of possible reasons for watching television, and asked to say whether or not each one applied to him. Table 64 shows the percentage of respondents in each sample saying that each of two reasons for viewing, both relating to the peer group, applied to them. Although the difference is not statistically significant, the number of control-sample boys saying they watched television because their friends would be watching was ten per cent higher than for the probationers. A very similar relationship holds for the girls, although the difference is again not statistically significant. However, the differences between the probationers and their matched controls do become statistically significant for both sexes when the data from a second, more pointed question are analysed.

After being asked whether they ever viewed 'because you know your friends will be watching', interviewees were invited to say whether they ever watched 'because you want to be able to talk to your friends about what's on'. Comparing the control sample with the probationer sample, 18 per cent more boys, and 27 per cent more girls said that this was a reason for them watching television. Table 65 shows that these differences are validated by the significant inter-sample differences between the frequencies with which probationers and control-sample members actually claimed to talk about television with their friends. In the case of both sexes, probationers scored significantly lower than the members of either of the two other samples.

Why did the probationers talk about television to their friends less often than members of the matched control sample, and place less importance on the possibility of holding conversations of this sort? Harking back to the argument about the differential availability of certain sorts of person advanced above, in connection with findings on the frequency with which members of different samples talked to their parents about television, it could be hypothesized that probationers merely had fewer friends than did those in the control sample, and that this is sufficient to account for the fact that probationers talked to friends significantly less often. Unfortunately, no data was collected on the actual *number* of friends claimed by sample members. However, data on how well members of the different samples got on with their peers suggests that there were *no* gross differences between samples in this respect.[3]

Probation officers, head teachers, youth club leaders and other

Table 64

Watching television for reasons relating to peer group

Reason	*Probationer sample*	*Control sample*	*Lower M.C sample*
	MALES		
Friends known to be watching	96 (34%)	43 (44%)	25 (20%)
Probationer sample and Control sample:		$Z = 1.72$; N.S.	
Probationer sample and Lower M.C. sample:		$Z = 2.97$; $p < 0.01$	
Control sample and Lower M.C. sample:		$Z = 3.92$; $p < 0.001$	
To be able to discuss programme with friends	185 (66%)	82 (84%)	63 (50%)
Probationer sample and Control sample:		$Z = 3.33$; $p < 0.001$	
Probationer sample and Lower M.C. sample:		$Z = 3.11$; $p < 0.01$	
Control sample and Lower M.C. sample:		$Z = 5.30$; $p < 0.001$	
	FEMALES		
Friends known to be watching	11 (21%)	16 (35%)	11 (19%)
Probationer sample and Control sample:		$Z = 1.56$; N.S.	
Probationer sample and Lower M.C. sample:		$Z = 0.24$; N.S.	
Control sample and Lower M.C. sample:		$Z = 1.83$; N.S.	
To be able to discuss programme with friends	26 (49%)	35 (76%)	36 (62%)
Probationer sample and Control sample:		$Z = 2.76$; $p < 0.01$	
Probationer sample and Lower M.C. sample:		$Z = 1.38$; N.S.	
Control sample and Lower M.C. sample:		$Z = 1.53$; N.S.	

Table 65

Frequency of conversation about television with friends

Probationer sample	*Control sample*		*Lower M.C. sample*
	MALES		
1.99	2.23		2.18
Probationer sample and Control sample:		$t = 2.19$; $p < 0.05$	
Probationer sample and Lower M.C. sample:		$t = 1.95$; N.S.	
Control sample and Lower M.C. sample:		$t = 0.50$; N.S.	
	FEMALES		
1.85	2.50		2.31
Probationer sample and Control sample:		$t = 3.64$; $p < 0.001$	
Probationer sample and Lower M.C. sample:		$t = 2.71$; $p < 0.01$	
Control sample and Lower M.C. sample:		$t = 1.32$; N.S.	

responsible adults were asked to rate each respondent on a five-point scale in terms of how well he or she got on with persons of his or her own sex. Scores were assigned to the five positions on the scale, with four points being given for a rating of 'Well above average' and 0 points for a score of 'Well below average'. Means for each group were calculated from these scores, and these averages are shown in Table 66. Middle-class respondents of both sexes scored significantly higher on average than members of either of the other two samples, while the means for the probationer and control samples were virtually identical for both boys and girls. It may of course be that the lower middle-class respondents received high ratings through the operation of a halo effect, so that adolescents whom adults approved of generally on account of their closer adherence to middle-class norms came to be perceived as doing better in other respects as well, including ease of social interaction with same-sex peers. But if such an effect *were* operating, it might be expected that the probationers, in a similar fashion, would have received *lower* ratings than were in fact justified. And the table gives no indication of this. There certainly seems no reason to suppose that the probationers received *higher* ratings than they in fact merited. Hence we must accept that they got on with their friends just as well as control-sample members. There thus seem to be no grounds for explaining the probationers' lower frequency of conversation about television with friends in terms of their inability to *find* friends and get on reasonably well with them.

Boys were asked how often they talked to their girlfriend about television, and girls how often they talked to their boyfriend. Table 67 shows that control-sample boys tended to talk about television to their girlfriends more than male probationers, although the difference was not statistically significant, while the lower middle-class boys scored on average highest of all, and significantly higher than both the two other samples. This pattern of differences is repeated in Table 68, which shows how well respondents got on with members of the opposite sex: here again, middle-class respondents did noticeably better than either of the other groups. Putting together the data on boys shown in these two tables, it is tempting to suggest that it was the ease with which boys got on with their girlfriends which determined how frequently they conversed about television.[4]

The three groups of girls did not differ among themselves either in the frequency with which they claimed to talk to boyfriends about television, or in the ease with which they got on with members of the opposite sex, as this was estimated by adults who knew them. It is however interesting to note that the greatest discrepancy between boys and girls on both of these variables occurred in the sample of probationers, with girls scoring higher than boys both in terms of

Table 66

Peer relationships with own sex

Probationer sample	*Control sample*	*Lower M.C. sample*
	MALES	
1.84	1.85	2.33
(N = 199)	(N = 96)	(N = 125)

Probationer sample and Control sample:	$t = 0.15$; N.S.
Probationer sample and Lower M.C. sample:	$t = 5.78$; $p < 0.001$
Control sample and Lower M.C. sample:	$t = 5.59$; $p < 0.001$

Probationer sample	*Control sample*	*Lower M.C. sample*
	FEMALES	
1.81	1.80	2.43
(N = 36)	(N = 45)	(N = 56)

Probationer sample and Control sample:	$t = 0.42$; N.S.
Probationer sample and Lower M.C. sample:	$t = 4.53$; $p < 0.001$
Control sample and Lower M.C. sample:	$t = 5.30$; $p < 0.001$

Table 67

Frequency of conversation about television with girlfriend/boyfriend

Probationer sample	*Control sample*	*Lower M.C. sample*
	MALES	
0.43	0.61	0.92

Probationer sample and Control sample:	$t = 1.77$; N.S.
Probationer sample and Lower M.C. sample:	$t = 5.07$; $p < 0.001$
Control sample and Lower M.C. sample:	$t = 2.29$; $p < 0.05$

Probationer sample	*Control sample*	*Lower M.C. sample*
	FEMALES	
0.79	0.63	0.84

Probationer sample and Control sample:	$t = 0.81$; N.S.
Probationer sample and Lower M.C. sample:	$t = 0.27$; N.S.
Control sample and Lower M.C. sample:	$t = 1.02$; N.S.

Table 68

Peer relationships with opposite sex

Probationer sample	*Control sample*	*Lower M.C. sample*
	MALES	
1.85	1.89	2.26

Probationer sample and Control sample:	$t = 0.29$; N.S.
Probationer sample and Lower M.C. sample:	$t = 0.90$; N.S.
Control sample and Lower M.C. sample:	$t = 1.37$; N.S.

Probationer sample	*Control sample*	*Lower M.C. sample*
	FEMALES	
2.31	2.13	2.13

Probationer sample and Control sample:	$t = 0.62$; N.S.
Probationer sample and Lower M.C. sample:	$t = 0.67$; N.S.
Control sample and Lower M.C. sample:	$t = 0.03$; N.S.

getting on with boyfriends and in terms of talking to them about television. It has been pointed out commonly enough that male delinquents tend to congregate in all-male groups (whether these are thought to deserve the label 'gangs' or not), and to reject any form of behaviour associated with the female role, while female delinquents' offences are often associated with sexual deviance. Such established differences between male and female delinquents may well be reflected in these data.

During the analysis of the survey findings, each respondent was assigned a score to represent the overall frequency with which he or she talked to other people about television. The maximum possible score on this index was 24. Table 69 shows the average score for each group. In general, probationers talked about television less than was the case with either of the other samples. The average score for male probationers was significantly below that for either control-sample males or lower middle-class males, while female probationers achieved an average score lower than that of girls in the two other groups, and significantly below that for the lower middle-class group.

Table 69

Overall frequency of conversation about television

Probationer sample	*Control sample*		*Lower M.C. sample*
	MALES		
6.73	8.48		8.04
Probationer sample and Control sample:		$t = 4.24$; $p < 0.001$	
Probationer sample and Lower M.C. sample:		$t = 3.59$; $p < 0.001$	
Control sample and Lower M.C. sample:		$t = 0.98$; N.S.	
	FEMALES		
7.40	8.50		8.81
Probationer sample and Control sample:		$t = 1.56$; N.S.	
Probationer sample and Lower M.C. sample:		$t = 2.16$; $p < 0.05$	
Control sample and Lower M.C. sample:		$t = 0.45$; N.S.	

In attempting to explain the fact that probationers of both sexes talked less about television to others than would be predicted on the basis of the scores achieved by the control sample, it might again be suggested that they just had fewer *opportunities* of talking to other people. Yet this does not seem to be the case. Chart 4 shows both how often, on average, members of each sample viewed television in company with other people, and how often, on average, they talked about television with others. Concentrating on the differences

CHART 4

Overall indices of viewing in company
and conversation about television
(data points from Tables 35 and 69)

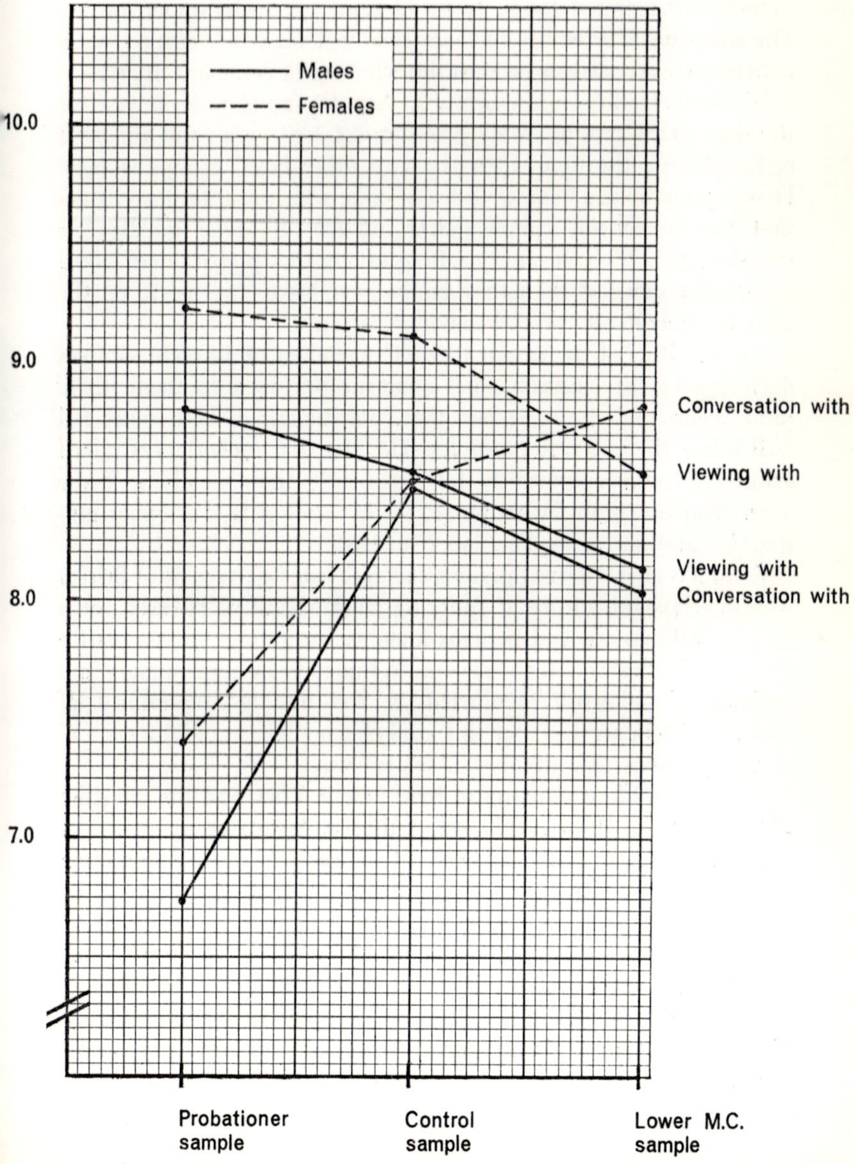

between the probationer sample and the control sample, it is clear that the members of the latter group viewed in company somewhat less than the probationers, but talked about television considerably more. If the frequency with which different groups viewed in company is taken as a rough index of the *opportunities* which were available for talking about what was seen, then it is clear that the probationers' lower scores on the measure of 'conversation about' *cannot* be accounted for in terms of restricted social opportunities. The situation is brought out more clearly if the 'conversation with' indices are expressed in ratio to the 'viewing in company' indices.

When these ratios are examined, they are found to be lowest for the delinquent groups. Male probationers achieve a ratio index of 0.76, whereas the figures for the two other boys' groups is 0.99. Female probationers have an index of 0.80, which is again lower than that for the control sample (0.93) and for the lower middle-class sample (1.03). We seem fairly safe in saying that probationers talked about television less than they might have been expected to on the basis of general adolescent behaviour patterns.

As indicators of the extent to which members of different samples made use of the conversational opportunities open to them, these ratios again underline the fact that probationers of both sexes fall well below the two other samples. The fact that probationers talked about television significantly less frequently than members of the other samples cannot be accounted for in terms of lack of opportunities. An alternative explanation is needed.

Klein has characterized persons exemplifying one variant of British working-class subculture as suffering from 'cognitive poverty', or a basic unwillingness, and perhaps inability, to indulge in speculative thought.[5] In the section she devotes to this topic in her summary and analysis of fieldwork conducted in the mining community of 'Ashton', Klein accounts for this kind of semi-deliberate limitation of mental horizons in a number of related ways.

For miners, life is a matter of hard practicalities, and the job itself calls for little in the way of abstract theory. Again, an acceptance of the social structure as it is leads to a downgrading of speculation about other possibilities, while the fear of disappointment and being 'let down' militates against day-dreaming even within the bounds of the given round of life. The close-knit nature of the community, and the limited range of social contacts which mining as a job imposes, both served to limit the need to practise other roles imaginatively. This in its turn militates against the contemplation of varied possibilities.

It seems possible to argue that some of the probationers studied here had been brought up in a similar atmosphere, though their social backgrounds clearly differ in a number of obvious ways from those of the miners referred to by Klein. The economic uncertainties associated with broken homes, or with homes in which the father performs the role of breadwinner poorly and intermittently, may equally encourage a limit to be placed on speculation about future possibilities. Again, a rejection of intellectual values may involve a rejection of the formal training in discursive thought which the school system can at its best offer, while uninterested, hostile or absent parents cannot be expected to provide the depth of emotional and mental stimulation usual in more stable domestic circumstances.

If the probationers studied here were hampered in these ways, then it is perhaps not surprising that they talked less to others about television than was the case with the adolescents in the control sample. Talking *about* television implies an ability to distance oneself from the programme content, to examine it critically, at least in a crude way, and to compare and contrast the experiences it offers with other ideas and other images of life. The probationers may to some extent have lacked both the skills and the range of social perspectives which make this sort of activity both possible and worthwhile. The argument becomes more striking when we remember that the delinquents were just as intelligent as their matched controls. In any event, this line of reasoning suggests that possibly the probationers did not merely talk *less* about television, but also conversed in a different *manner*. Other data would seem to bear out this contention.

Kinds of reasons

Bernstein has suggested that members of the working class are often limited in their speech behaviour to what he terms a 'restricted code'.[6] By this he means a form of conversation which is not merely or most importantly limited in terms of its richness of vocabulary, but is also relatively poor in the range of syntactical constructions employed: compound sentences are common (" . . . and . . . and . . . ") but the use of dependent and relative clauses is rare. Klein, in her discussion of Bernstein's thinking as it relates to child-rearing practices, notes in addition that one of the characteristics of social dialects of this sort is a "simple, repetitive and often inaccurate use of conjunctions" such as 'because', while, in addition, "statements of fact [are] often used as both reason and conclusion, or more accurately, reason and conclusion are confounded to produce a categoric statement".[7]

In its crudest form, this kind of statement might be represented by a mother urging her child to obey a command with the admonition "Do it because I'm your mother", or "Do it because I'm telling you to." In other words, particular reasons are not advanced, and the consequences likely to result from different patterns of action are not referred to.

Perhaps the probationers' relative lack of verbal skill, particularly skill in the discursive elaboration of 'because' statements, is revealed by the data shown in Table 70 and Table 71. When asked to say why they liked the particular programme favourites they had listed, a number of respondents merely reiterated the name for the category to which the programme belonged. For example, when asked to say why he liked *Top of the Pops*, a respondent might say "Because it's a pop programme." Table 70 shows that this sort of 'reason' for liking a particular programme was considerably more common among probationers of both sexes than it was among members of the control sample or members of the middle-class sample. In terms of cognitive development, this is of course a very primitive sort of reason to give, but one which might well be characteristic of speakers relying on a restricted code.

As Table 72 shows, probationers were far less likely than members of the other samples to refer to television production techniques when answering the same question. This is again the sort of analytical response one might expect from those who are both able to distance themselves more from the object of discourse and in a

Table 70

Descriptions of programme type as reason for liking favourite programme

Probationer sample	*Control sample*	*Lower M.C. sample*
	MALES	
316 (42%)	51 (17%)	62 (15%)
(N = 757)	(N = 294)	(N = 407)

Probationer sample and Control sample: $Z = 7.44$; $p < 0.001$
Probationer sample and Lower M.C. sample: $Z = 9.20$; $p < 0.001$
Control sample and Lower M.C. sample: $Z = 0.75$; N.S.

Probationer sample	*Control sample*	*Lower M.C. sample*
	FEMALES	
76 (49%)	14 (9%)	20 (11%)
(N = 156)	(N = 160)	(N = 188)

Probationer sample and Control sample: $Z = 7.87$; $p < 0.001$
Probationer sample and Lower M.C. sample: $Z = 7.84$; $p < 0.001$
Control sample and Lower M.C. sample: $Z = 0.59$; N.S.

Table 71

References to production techniques as reason for liking favourite programme

Probationer sample	*Control sample*	*Lower M.C. sample*
	MALES	
70 (9%)	111 (38%)	230 (57%)
(N = 757)	(N = 294)	(N = 407)

Probationer sample and Control sample: $Z = 10.99$; $p < 0.001$
Probationer sample and Lower M.C. sample: $Z = 17.58$; $p < 0.001$
Control sample and Lower M.C. sample: $Z = 4.90$; $p < 0.001$

	FEMALES	
18 (12%)	79 (49%)	96 (51%)
(N = 156)	(N = 160)	(N = 188)

Probationer sample and Control sample: $Z = 7.29$; $p < 0.001$
Probationer sample and Lower M.C. sample: $Z = 7.75$; $p < 0.001$
Control sample and Lower M.C. sample: $Z = 0.31$; N.S.

position to practise this sort of evaluation in conversation with peers and adults.

One further set of findings lend some additional support to the idea that male probationers lacked the cognitive and imaginative flexibility characteristic of the other samples. Before television had been mentioned at all, all respondents were invited to say who they would like to be, if they were not themselves, and had the chance to

Table 72

Ability to provide general identification choice

Choice given	*Probationer sample*	*Control sample*	*Lower M.C. sample*
		MALES	
Yes	183 (65%)	76 (78%)	108 (85%)
No	98 (35%)	22 (22%)	19 (15%)
Total	281 (100%)	98 (100%)	127 (100%)

Probationer sample and Control sample: $\chi^2 = 4.63$; $p < 0.05$
Probationer sample and Lower M.C. sample: $\chi^2 = 16.00$; $p < 0.001$
Control sample and Lower M.C. sample: $\chi^2 = 1.61$; N.S.

		FEMALES	
Yes	41 (77%)	33 (72%)	46 (79%)
No	12 (23%)	13 (28%)	12 (21%)
Total	53 (100%)	46 (100%)	58 (100%)

Probationer sample and Control sample: $\chi^2 = 0.17$; N.S.
Probationer sample and Lower M.C. sample: $\chi^2 = 0.00$; N.S.
Control sample and Lower M.C. sample: $\chi^2 = 0.44$; N.S.

be someone else. Table 72 shows that about three-quarters of all girls were able to give an answer to this question, and that the three groups of girls did not differ significantly in this respect. So far as boys were concerned, however, the proportion unable to say whom they would like to be was highest among the probationers, and lowest among the middle-class boys. There was a statistically significant difference here between the probationer and control samples.

While the fact that the members of the middle-class sample were, on average, significantly more intelligent than the members of the control sample might account for the fact that the former group were more readily able to name somebody they would like to be, this factor cannot of course account for so many male probationers being unable to name their choice. This situation might, however, be explicable in terms of the probationers' lack of role-playing skills.

Klein has suggested that a lack of ability in role-playing is one aspect of cognitive poverty.[8] If the male probationers' style of thought is characterizable in terms of cognitive poverty, the fact that they were less ready to mention somebody they would like to be may be explicable in terms of an attendant lack of skill in role-playing.

But we must now turn to consider the possible consequences of the differences reported in this chapter. Probationers seem to have talked less about television to other people, and to have been less able to articulate their reactions to the medium. Why might this be important?

Briefly put, it seems possible that those who discuss their media experience less will be more subject to media influence. Those who discuss what they have seen with others will be led to evaluate the media content involved, and come to a more considered judgment about it. Conversely, those who watch television but never or seldom talk about it may accept its suggestions in a far less critical fashion. Mass communications researchers have come to realize that the media audience is characteristically not made up of isolated individuals, but of social persons with membership in a wide range of formal and informal groups. More important, these groups serve in effect to shield the audience member from the direct effects of the media, since he is subject not merely to impersonal forces, but to pressures from all the other people in his environment as well.[9] The pressures he is under from these other people may be quite different from the pressures exerted by media content, though of course in some cases group influences may reinforce media effects, rather than weaken them. In the case of the delinquents studied here, it seems reasonable to suggest that media influence was likely to have been less modified by group influences than was the case with members of the non-probationer samples, *even if* group pressures were working in the same direction as media pressures. For group pressures can only be transmitted through conversation, while the probationers seem to have talked less about television than did the other adolescents.

VII

MEDIA IDENTIFICATIONS AND 'POP' MUSIC

General identification choices

We noted in the last chapter that probationers were apparently less frequently able to name some person they would have liked to be than were members of the other samples. An analysis of the answers which *were* given to the initial identification question is shown in Table 73. So far as boys are concerned, breaking down all responses into the six listed categories leads to distributions for the probationer and control samples which are not significantly different. Lower

Table 73
General identification choices

Type of choice	*Probationer sample*	*Control sample*	*Lower M.C. sample*
	MALES		
Television personality or character	14 (9%)	8 (12%)	11 (12%)
Figure connected with other medium	20 (13%)	5 (7%)	6 (6%)
'Pop' musician	25 (16%)	13 (19%)	15 (16%)
Sports personality	31 (20%)	5 (7%)	33 (35%)
Occupation	20 (13%)	10 (15%)	10 (11%)
Other	46 (29%)	26 (39%)	20 (21%)
Total	156 (100%)	67 (100%)	95 (100%)

Probationer sample and Control sample: $\chi^2 = 7.82$; N.S.
Probationer sample and Lower M.C. sample: $\chi^2 = 9.79$; N.S.
Control sample and Lower M.C. sample: $\chi^2 = 17.82$; $p < 0.01$

Type of choice	*Probationer sample*	*Control sample*	*Lower M.C. sample*
	FEMALES		
Mass media personality or character	7 (21%)	16 (50%)	15 (44%)
'Pop' musician	13 (38%)	10 (31%)	6 (18%)
Other	14 (41%)	6 (19%)	13 (38%)
Total	34 (100%)	32 (100%)	34 (100%)

Probationer sample and Control sample: $\chi^2 = 7.06$; $p < 0.05$
Probationer sample and Lower M.C. sample: $\chi^2 = 5.53$; N.S.
Control sample and Lower M.C. sample: $\chi^2 = 3.55$; N.S.

Table 74

Differences between sexes in general identification choices (probationer sample)

Type of choice	*Males*	*Females*
Mass media personality or character	34 (22%)	7 (21%)
'Pop' musician	25 (16%)	13 (38%)
Other	97 (62%)	14 (41%)
Total	156 (100%)	34 (100%)

$\chi^2 = 9.01$; $p < 0.05$

middle-class boys differ from control-sample boys due mainly to the fact that surprisingly few control-sample males said that they would like to be someone famous in the sporting world. The smaller number of girls in each sample meant that fewer analytic categories could be used. Employing those shown in the table, a significant difference is revealed between the probationers and the control-sample girls: while more probationers mentioned a 'pop' musician, more control-sample girls mentioned other sorts of mass media personality or character. Within each sample, more girls than boys mentioned that they would have liked to be a 'pop' musician.

This inter-sex difference is particularly marked for the probationer sample, with a discrepancy of 22 per cent between boys and girls. Table 74 shows that it was the girls' more frequently stated desire to be a 'pop' musician that led to a statistically significant difference between the sexes.

Identification with media figures

During a later phase of the interview, respondents were asked again whom they would like to be, but this time the words 'on television' were introduced to delimit and focus the field of choice. The proportion of boys saying they would like to be a 'pop' singer falls as one moves from the probationer sample to the control sample. This series of differences is largely responsible for the distribution of choices between categories being significantly different when these two extreme samples are compared (Table 75). Within all samples, again more girls than boys wanted to be 'pop' singers, and there is a similar falling off in the proportion making this choice from probationer girls to middle-class girls. Here, the distribution of female probationers' responses is significantly different from that of either of the two other samples. Although among probationers of both sexes a higher proportion of respondents chose to be a 'pop' singer than was true of either of the other samples, there was still a highly significant difference between the sexes, as shown in Table 76. In terms of the salience of the world of 'pop' music, it will be recalled that Table 74 presents a very similar picture.

Besides being asked to give identification choices, respondents were also asked to name persons on television whom they regarded as their favourites. The responses could again be categorized into the sorts of classes already employed in the present series of tables. The distribution of male probationers' responses was virtually identical to that given by the control sample, though there was a significant difference between the control and middle-class samples. These findings are shown in Table 77. With girls, two of the three inter-sample differences are statistically significant. In broad terms, the table shows that this is due to the proportion of mentions of heroes or heroines increasing as one goes from the probationers to the middle-class sample, while the proportion of mentions of 'pop' singers once again falls off. And once again too, within the probationer sample, the ratio of girls' mentions of 'pop' singers to the number of mentions by boys is higher than for either of the two other samples, with girls giving

Table 75

Identification with persons on television

Category of person	*Probationer sample*	*Control sample*	*Lower M.C. sample*
	MALES		
Hero or heroine	98 (43%)	38 (48%)	56 (53%)
'Pop' singer	39 (17%)	9 (11%)	4 (4%)
Comedian, disc jockey, compère	36 (16%)	16 (20%)	21 (20%)
Other	57 (25%)	17 (21%)	25 (24%)
Total	230 (100%)	80 (100%)	106 (100%)

Probationer sample and Control sample: $\chi^2 = 2.55$; N.S.
Probationer sample and Lower M.C. sample: $\chi^2 = 12.29$; $p < 0.01$
Control sample and Lower M.C. sample: $\chi^2 = 4.01$; N.S.

Category of person	*Probationer sample*	*Control sample*	*Lower M.C. sample*
	FEMALES		
Heroine or hero	9 (25%)	14 (41%)	9 (26%)
'Pop' singer	22 (61%)	8 (24%)	7 (21%)
Other	5 (14%)	12 (35%)	18 (53%)
Total	36 (100%)	34 (100%)	34 (100%)

Probationer sample and Control sample: $\chi^2 = 10.45$; $p < 0.01$
Probationer sample and Lower M.C. sample: $\chi^2 = 15.06$; $p < 0.001$
Control sample and Lower M.C. sample: $\chi^2 = 2.35$; N.S.

Table 76

Sex differences in identification with persons on television (probationer sample)

Category of person	*Males*	*Females*
Hero or heroine	98 (43%)	9 (25%)
'Pop' singer	39 (17%)	22 (61%)
Other	93 (40%)	5 (14%)
Total	230 (100%)	36 (100%)

$\chi^2 = 34.82$; $p < 0.001$

Table 77

Favourite persons on television

Type of person	*Probationer sample*	*Control sample*	*Lower M.C. sample*
MALES			
Hero or heroine	220 (34%)	76 (32%)	115 (41%)
'Pop' singer	110 (17%)	42 (18%)	28 (10%)
Character in series or serial	65 (10%)	36 (15%)	28 (10%)
Comedian, disc jockey, compère	153 (24%)	55 (24%)	74 (26%)
Actor	54 (8%)	18 (8%)	22 (8%)
Other	42 (7%)	7 (3%)	13 (5%)
Total	644 (100%)	234 (100%)	280 (100%)

Probationer sample and Control sample: $\chi^2 = 8.33$; N.S.
Probationer sample and Lower M.C. sample: $\chi^2 = 10.91$; N.S.
Control sample and Lower M.C. sample: $\chi^2 = 12.75$; $p < 0.05$

FEMALES			
Heroine or hero	35 (23%)	54 (47%)	26 (18%)
'Pop' singer	63 (42%)	33 (29%)	74 (50%)
Character in series or serial	24 (16%)	17 (15%)	27 (18%)
Comedian, disc jockey, compère	18 (12%)	5 (4%)	20 (14%)
Other	10 (7%)	6 (5%)	1 (1%)
Total	150 (100%)	115 (100%)	148 (100%)

Probationer sample and Control sample: $\chi^2 = 18.68$; $p < 0.001$
Probationer sample and Lower M.C. sample: $\chi^2 = 36.97$; $p < 0.001$
Control sample and Lower M.C. sample: $\chi^2 = 5.30$; df = 3; N.S.

this sort of person as their favourite three times as often as boys. The difference between the sexes within the probationer sample is once again highly significant, as shown in Table 78.

Table 78

Sex differences in favourite person on television (probationer sample)

Type of person	*Males*	*Females*
Hero or heroine	220 (34%)	35 (23%)
'Pop' singer	110 (17%)	63 (42%)
Character in series or serial	65 (10%)	24 (16%)
Comedian, disc jockey, compère	153 (24%)	18 (12%)
Other	96 (15%)	10 (7%)
Total	644 (100%)	150 (100%)

$\chi^2 = 56.90$; $p < 0.001$

It is interesting that 'pop' singers featured strongly among the kinds of television personalities respondents particularly *disliked*. Table 79 shows that a familiar pattern is repeated. References to 'pop' singers are commonest amongst the probationers, and least common among the members of the middle-class group, while for each sample this type of reference is approximately twice as common among girls as among boys. The fact that both boys' and girls' groups are apparently arranged in this way along a continuum accounts for the fact that a comparison between the probationer and middle-class samples yields a statistically significant difference for both sexes.

'Pop' singers feature particularly prominently among both the likes *and* the dislikes of probationers of both sexes, although girls referred to 'pop' singers more often than boys. This suggests that the 'pop' music world had a particular salience for probationers. It also suggests that it was a world within which they were willing and able to make discriminations, knowing which styles they rejected as well as those they admired. Even so, the particular salience which 'pop' singers achieved among the dislikes of female probationers largely accounts for the significant inter-sex difference shown in Table 80.

In brief, sex and the sample they belonged to affected the salience which 'pop' music had for these adolescents. The world of 'pop' music was clearly of greater importance for girls than for boys, and of greater importance too for members of the probationer sample as compared with members of the two other groups. These two variables appear to interact so as to give the 'pop' world exceptionally high prominence in the eyes of the girls in the probationer sample.

The sex difference may possibly be due to the status of female adolescents encouraging a more restricted range of role-playing than

Table 79

Disliked persons on television

Type of person	*Probationer sample*	*Control sample*	*Lower M.C. sample*
	MALES		
Hero or heroine	19 (6%)	8 (7%)	12 (8%)
'Pop' singer	51 (17%)	11 (10%)	9 (6%)
Character in series or serial	64 (21%)	27 (24%)	28 (18%)
Comedian, disc jockey, compère	107 (35%)	46 (41%)	58 (37%)
Politician	29 (10%)	16 (14%)	29 (18%)
Other	33 (11%)	5 (4%)	22 (14%)
Total	303 (100%)	113 (100%)	158 (100)%

Probationer sample and Control sample: $\chi^2 = 9.17$; N.S.
Probationer sample and Lower M.C. sample: $\chi^2 = 17.99$; $p < 0.01$
Control sample and Lower M.C. sample: $\chi^2 = 9.66$; N.S.

Type of person	*Probationer sample*	*Control sample*	*Lower M.C. sample*
	FEMALES		
'Pop' singer	29 (32%)	11 (21%)	9 (12%)
Character in series or serial	19 (21%)	12 (23%)	17 (23%)
Comedian, disc jockey, compère	28 (31%)	18 (35%)	27 (37%)
Politician	3 (3%)	5 (10%)	13 (18%)
Other	12 (13%)	6 (12%)	7 (10%)
Total	91 (100%)	52 (100%)	73 (100%)

Probationer sample and Control sample: $\chi^2 = 1.98$; df = 3; N.S.
Probationer sample and Lower M.C. sample: $\chi^2 = 16.44$; $p < 0.01$
Control sample and Lower M.C. sample: $\chi^2 = 3.05$; N.S.

Table 80

Sex differences in disliked persons on television (probationer sample)

Type of person	*Males*	*Females*
'Pop' singer	51 (17%)	29 (32%)
Character in series or serial	64 (21%)	19 (21%)
Comedian, disc jockey, compère	107 (35%)	28 (31%)
Politician	29 (10%)	3 (3%)
Other	52 (17%)	12 (13%)
Total	303 (100%)	91 (100%)

$\chi^2 = 12.29$; $p < 0.05$

is found with boys. This in its turn is simply because the range of both social and occupational roles *is* still less wide for girls than boys. Since girls are *less* likely to imagine themselves in a number of the occupational roles depicted on television, perhaps they are *more* likely to name choices from the 'pop' music world. After all, this milieu provides many examples of girls of roughly their own age filling the roles of both performer and audience member.

Leaving aside the sex differences, the inter-sample differences only seem explicable in terms of the different meanings which the world of 'pop' music can have for individuals with differing emotional needs and differing views of their social situation. The 'pop' music business is in fact closely integrated into the rest of the entertainment industry. Nevertheless, its performers and products can easily be perceived as standing in various sorts of opposition to the adult world, to traditional standards, and to established authority. Indeed, particular groups and performers have clearly specialized in projecting a consciously oppositional 'image'. The probationers in this study may well have attached this last kind of meaning to the world of 'pop' music.

Sugarman, in his examination of the relationships between male adolescents' attitudes towards the school and their involvement in youth culture, points towards a more theoretical account of this kind of situation.[1] With reference to prior theory and research, Sugarman notes that the idea of an adolescent culture in value conflict with the adult world has been mooted quite frequently. However, past research also shows that there are considerable class differences in adolescents' attachment to and involvement in youth culture and that it is not simply the adult world in general which adolescents may reject, but rather the values of the disciplined, middle-class world. Sugarman's own data suggest a positive correlation between involvement in youth culture and rejection of the middle-class norms of the secondary school. We might therefore have expected the probationers to be particularly involved in cultural materials such as 'pop' music which currently constitute the most salient expressive aspect of youth culture. Sugarman notes that the appeal of 'pop' singers may partly lie in the fact that they have achieved success without having to pass examinations or achieve status gradually through seniority. If our probationers were indeed over-rejecters of middle-class work and career norms, then this aspect of the 'pop' milieu may have been particularly appealing to them. While for most adolescents interest in 'pop' music may be expressive of a temporary adolescent status and of a temporary (and at most partial) rejection of adult norms, for delinquents it may become expressive of a more permanent status where ordinary avenues to success are seen as blocked and where the attendant middle-class norms are over-rejected.

It is worth pointing out, too, that youth culture as studied in the United States and Britain places no stress on academic achievement. It is again only a small step to a position where such achievement is actively rejected. As we have seen, the probationers were at the least *under*-users of television for knowledge and information, even if our data do not allow us to show that they were actually hostile to the relevant kinds of television programmes.

What we are noting here, in brief, is that there may be a natural affinity between the values and orientations expressed in 'pop' music and related cultural materials and the values and orientations characteristic of delinquents.[2] By emphasizing certain themes in youth culture, adolescent delinquents may find a means of expressing their own more particular status. Sugarman writes that "Youth culture . . . is . . . the culture of the non-mobile working-class, the downwardly-mobile and of those who cherish hopes of mobility along channels where the criteria of school do not apply."[3] Although delinquency is not usually seen as a way of achieving mobility, even by delinquents, it is potentially a means of achieving status and success within a particular social stratum.

170 – 181

VIII

CONCLUSION

The findings

On the basis of the findings reported in the preceding chapters, it seems clear that the juvenile delinquents studied did differ from the matched control sample in their pattern of television-viewing behaviour. However, the differences are more clear-cut, and more easily explicable, in the case of boys. Within the context of the present study, this is of course in some measure due to the smaller numbers of girls included in the samples. Yet it must also be said that because far fewer girls are legally classified as delinquent, less is known about their characteristic behaviour patterns, and sociological theorizing about the aetiology of female delinquency is much less advanced than it is for males.

But although there are a number of statistically significant inter-sample differences, these have mostly to do with the *uses* to which television viewing was put, rather than being differences in the gross *amount* of exposure. Indeed, the available data suggest too that there

Table 81

Ability to mention favourite programmes

	Probationer sample	*Control sample*	*Lower M.C. sample*
	MALES		
Maximum possible number of mentions	1686	588	762
Actual number of mentions	1290 (76.5%)	471 (80.1%)	618 (81.1%)

Probationer sample and Control sample: $Z = 1.79$; N.S.
Probationer sample and Lower M.C. sample: $Z = 2.54$; $p < 0.05$
Control sample and Lower M.C. sample: $Z = 0.46$; N.S.

	Probationer sample	*Control sample*	*Lower M.C. sample*
	FEMALES		
Maximum possible number of mentions	318	276	348
Actual number of mentions	260 (81.8%)	223 (80.8%)	304 (87.4%)

Probationer sample and Control sample: $Z = 0.30$; N.S.
Probationer sample and Lower M.C. sample: $Z = 2.00$; $p < 0.05$
Control sample and Lower M.C. sample: $Z = 2.25$; $p < 0.05$

Table 82
Favourite programmes

Programme	*Probationer sample* R	(F)	*Control sample* R	(F)	*Lower M.C. sample* R	(F)
		MALES				
The Man from UNCLE	1	(94)	1	(38)	1	(75)
Top of the Pops	2	(69)	3	(27)	2	(40)
Coronation Street	3	(56)	5	(17)	7.5	(8)
Bonanza	4	(51)	4	(18)	9.5	(6)
Crossroads	5	(44)	2	(29)	4.5	(14)
The Avengers	6.5	(30)	6	(15)	3	(32)
Mystery and Imagination	6.5	(30)	8	(9)	6	(9)
The Saint	8	(27)	7	(10)	7.5	(8)
The Big Valley	9.5	(22)	9	(8)	9.5	(6)
Danger Man	9.5	(22)	10	(7)	4.5	(14)
		FEMALES				
Top of the Pops	1	(33)	2	(31)	3	(29)
Coronation Street	2	(27)	4	(15)	4	(14)
Crossroads	3	(25)	3	(20)	2	(32)
The Man from UNCLE	4	(13)	1	(32)	1	(34)
Emergency Ward 10	5	(11)	6	(4)	6	(7)
The Avengers	6	(5)	5	(11)	5	(12)

were few differences in the actual *programmes* viewed. Members of the lower middle-class sample were able to mention rather more favourite programmes than was the case with the other samples (Table 81). However, Table 82 shows that there were only slight differences in the salience which particular programmes achieved in the lists of favourites produced by each sample. The three samples differed in terms of how they perceived and used television programmes rather than in terms of which programmes they commonly viewed or listed as favourites. We first review here what appear to be some of the more theoretically important inter-sample differences in television usage, before relating the patterns characteristic of the probationers more generally than has so far been done to the kind of backgrounds from which most of them came.

As Sections IV and V have shown, the three samples of boys differed importantly in the degree of importance they attached to exciting and informative content. Data from these sections are brought together in Table 83. As compared with their matched controls, probationers expressed a disproportionate interest in what we have termed 'Exciting' programmes. As the table shows, the inter-sample ranking is replicated from three completely independent sets of questions. Put mathematically, the chance of discovering this particular set of parallel rankings is 1 in 216. We believe it represents a genuine set of differences.

In Section VI, data were reported which suggested that the probationers talked to other people about their television viewing considerably less than was the case with the two other samples. In addition, the probationers seem to have been less able to articulate their reactions to what they had seen. These differences seem to us to constitute the second most important way in which the probationers were unlike the members of the control sample.

Section VI also showed that the probationers were somewhat less ready to name identification figures than members of the other samples. In Section IV we saw too that the probationers were particularly aware of the hero figures who appear in television programmes. In addition, data reported in Section VII showed probationers of both sexes, but particularly girls, to be more likely to mention 'pop' performers than members of the other samples.

Table 83

*Differences in the uses of television**

Index	*Probationer sample*	*Control sample*	*Lower M.C. sample*
Preferences for kinds of programmes			
Ratio of 'Exciting' kinds to 'Educational and cultural' kinds	5.24	2.63	1.52
Descriptions of ideal television fare			
Ratio of 'Exciting'† to 'Interesting'‡	3.05	2.11	0.61
Feelings ideally produced by television			
Ratio of 'Excitement' to 'Interest'	1.00	0.64	0.25

* For original data see Tables 38, 40, 41, 51, 53 and 55.

† Includes 'Thrilling' and 'Adventurous'.

‡ Includes 'Informative'.

Theoretical explanations

A major difficulty in offering satisfactory *post hoc* explanations of the various inter-sample differences discovered in the present investigation stems from the very multiplicity of available theories of delinquency. Focusing for the moment merely on subcultural theories, it is worth mentioning that as recently as 1966 Downes could point to a number of unresolved ambiguities in American theorizing, and go on to emphasize "the paucity of English work on the sociology of crime".[1] In this country, the situation has not changed radically over the last four years. In addition, of course, the applicability of American approaches to the British situation has to be carefully scrutinized, as Downes himself points out. In terms of the present enquiry, itself merely exploratory, the lack of both adequate data and well-tested theory has meant that the explanatory paragraphs included above have been somewhat eclectic in their use of prior research into delinquency.

One particular warning needs to be issued. The substantive sections of this report imply that there is but one type of delinquent, and that once the dynamics of 'delinquency' have been understood then the characteristic patterns of delinquents' media usage will be readily explicable in terms of a unitary theory. However, such a simple view of the matter has become less and less tenable as research into delinquency has moved forward. It could well be that the probationers studied here comprised a number of different types of delinquent, each having a specific pattern of media usage, and each pattern calling for a somewhat different theoretical explanation. The situation may well be more complex than the data as presented here suggest. For example, 17 of the male probationers had fathers in non-manual employment categories, and it seems likely that subcultural accounts of the media usage patterns of the remainder of the probationer sample would only be applicable to these cases in a severely modified form.

Testing the relative merits of different theories of juvenile delinquency was not, of course, among the objectives of this piece of exploratory research. Yet the types of differences which are established here between the probationers and the members of the control sample do suggest the appropriateness of one type of subcultural theory. It is perhaps worthwhile reorganizing some of the findings presented above so that the ways in which they can be related back to a particular life-style become more readily apparent. The discussion will be limited to boys, since the presentation above has revealed a number of differences between boys and girls. The lack of success in matching also makes valid comparisons between the girls' groups more difficult.

The background data on the male probationers suggests that they came in the main from socially disorganized working-class families. Their fathers tended to be in semi-skilled or unskilled trades, and to have poor employment records. One or both natural parents were commonly missing from the home; where present in the home they often suffered from ill health. Respondents had a strikingly high number of siblings, and the households were in general large. Given these indications, it seems likely that probationers came from homes where there were other deficiencies too. These may have included a haphazard system of rewards and punishments during early socialization, an attenuated degree of attention from both parents, in terms of both affection and interest (partially the direct result of the number of children in the home), and in general a lack of forward planning in the management of domestic affairs.

In this sort of milieu, children are likely to receive less than the normal share of what might be termed cognitive training, even for the lower working-class stratum. Life is frequently a matter of living from hand to mouth, with fathers in low-paid and irregular employment, remarkably few mothers working, and one or both parents in poor health: thus thinking about the future becomes of little importance or point. But this in its turn means that offspring may receive little training or encouragement in thinking about sequences of events in a logically connected fashion. With parents who have themselves enjoyed little formal education, and see little point in it for their children, the home may well be marked by a paucity of conversation which goes beyond the immediate registration of emotional states of affairs. The lack of facility in an elaborated verbal code will itself hamper the development of any ability to foresee and plan the future, and to relate events by means of causal chains. The training of children is likely to be conducted in terms of peremptory commands, rather than by means of argument and explanation, and this again works against a child developing a habit of imagining the end result of planned activity. In households of this sort, boys may see their fathers as failures and be little drawn to emulation of them, while the large world of middle-class institutions (including the school) is at best alien and at worst totally rejected. Social horizons will in general be limited, and children will have restricted opportunities to mix with a range of persons and thus appreciate a range of social viewpoints.

This is of course an 'ideal type' description, and by no means all of the male probationers exemplified all of these background characteristics. Nevertheless, what the survey revealed as characteristic of the probationers' attitudes and behaviour, including mainly data on the school and television, fits in quite well with this general picture of social deprivation and dislocation.

Although the boys were bright enough, they were absent from

school a good deal, and were scored rather low in terms of school attainment and conduct. Whether the boys actually rejected the values of the school (so that a reaction-formation theory becomes appropriate), or were merely apathetic to them, this seems to be reflected in their lack of interest in, or hostility to, educational material on television. The probationers' upbringing had no doubt provided little home training in the development of cognitive skills, so that formal education became a sphere where little in the way of success was to be expected. Further, lack of awareness of any utility in conscious thinking about the future or in forward planning perhaps meant that the merely utilitarian value of education had never really dawned on them. And the way conversational topics were handled in the home may have meant that the notion of talking 'about' television to others (and about other matters too) was a relatively unfamiliar one for the probationers. This particular aspect of the probationers' lack of cognitive skills seems to have been reflected too in their relative inability to say *why* they liked particular programmes.

But the fact that probationers talked about television to others less than did matched controls perhaps had other, contributory causes. For on various counts it can be suggested that the probationers were deficient in social, as well as cognitive, skills. Although they commonly watched in the company of a number of other persons, this was not the case with girlfriends, perhaps suggesting that they lacked the social skills necessary to get on successfully with members of the opposite sex. Again, their general home backgrounds may have failed to offer much training in skills of this sort, and specifically in putting themselves in the other person's place. Perhaps this accounts for the fact that probationers were less readily able to say who they would have liked to be if they were not themselves. But of course, an alternative explanation of the relative lack of girlfriends among the probationers may be due to the fact that they were attempting to emphasize their masculinity by moving in all-male groups.

Certainly, these boys' apparently greater liking for the heroes seen on television may have been due to an over-emphasis on masculinity produced through a process of reaction-formation (and untrammelled by parental control), or merely to the search for more satisfying male identification figures than their fathers provided. And perhaps this explains in part their greater liking for programmes featuring adventurous heroes.

The probationers' greater expressed liking for excitement-producing programmes may also be related to the greater importance they attached to series and serials featuring this kind of character. However, the relatively little stress which these boys placed on television as a form of relaxation may mean that television was looked to as a tension-producing instrument quite apart from the super-masculine behaviour which programmes commonly portray. It may be that the probationers needed to express themselves through both real and fantasy forms of 'exploit' merely because they had not been trained to use other sorts of emotional outlet. However, this does not seem a very satisfactory sort of explanation.

Schramm and his co-authors have argued that the heavy viewing of television, and more particularly the heavy viewing of purely entertainment programmes, commonly goes together with the seeking of immediate rewards in other areas of life, whereas relatively heavy use of television's informational content, and heavy reliance on printed media, go together with patterns of delayed or deferred gratification which again are repeated in the other realms of behaviour.[2] In terms of television, they refer to the first kind of orientation as 'fantasy seeking', and to the second as 'reality seeking'. It is clear that our own data, particularly those summarized in Table 83, could be discussed in similar terms. As one goes from the probationers to the lower middle-class respondents, fantasy seeking decreases and reality seeking increases; in terms of television, the probationers scored highest on immediate gratification, the lower middle-class respondents highest on deferred gratification. But while the terms utilized by Schramm and his team are useful in themselves, and allow us to link the present findings to a further body of prior research, they perhaps have an added value as well. We have already suggested that the probationers came from a lower working-class stratum probably marked by the lack of any forward planning or orientation to the future. The notion of immediate gratification in television viewing may indicate another way in which the probationers' patterns of media usage fitted in with their general orientation to life.

It would be wrong, however, to lay too much stress on the uniqueness of the probationers' television-viewing behaviour. As many of the tables show, in the majority of cases there were no differences, or only very slight differences, between the probationers and their controls. In the bulk of the tables, it is the members of the third, lower middle-class sample who are noticeably different from the two other groups. The major features of the delinquents' viewing behaviour must therefore commonly be explained in terms of their general social class background rather than in terms of the fact that they had been legally categorized as lawbreakers. Up to this point, explanatory comments have been directed to the probationers, rather than to both groups of mainly working-class respondents taken together. Yet in some instances a better account might perhaps be given by focusing first on the largely similar patterns of media behaviour here found to be characteristic of both the probationers *and* their controls. The particular features of the *delinquents'* tastes and choices could then perhaps be better explained as variations on a common theme, or intensifications of particular tendencies found also among the control-sample members.

This approach to the interpretation of the data presented above is suggested, in fact, by some of the available theoretical writing on delinquency itself. If we assume that juvenile delinquency is characteristically a working-class rather than middle-class kind of behaviour, then we obviously need to ask what it is about working-class life styles that produces this class-linked pattern of incidence. Miller has supplied one of the best known sets of answers to this question, and his characterization of working-class culture has particular relevance for the explanation of certain of the findings reported above.[3] Since we are not concerned here with theories of delinquency, a full rehearsal of Miller's position would be largely redundant. Attention will be drawn only to some of the more specifically relevant features of his thesis.

Miller provides a list of the 'focal concerns' of lower-class culture, and among these is the ideal of masculine toughness, with attendant notions of physical prowess, skill and fearlessness. Given these concerns, one would not be surprised to find the particular salience which the hero figures of television achieved among the delinquents in the present study, as compared with the controls and more middle-class boys. Here was one area where the probationers *were* distinctively different according to three separate indices. The second pertinent focal concern listed by Miller is centred on the goals of excitement, thrill, danger and risk. As we have seen, the probationers studied here seemed to look for material on television which was able to provide gratifications of these sorts. However, the control-sample members also showed interest in these sorts of material, and to a significantly greater extent than the lower middle-class respondents. Such tastes, in other words, seem partly a function of social class membership and partly a function of delinquent status: being a delinquent intensifies tastes which are also a more general feature of the working-class subculture. Miller's suggestions, in other words, should serve to remind us that we are faced with the task of explaining the viewing behaviour not just of juvenile delinquents, but of characteristically working-class juvenile delinquents, so that we must be conscious of the differences between the control and lower middle-class samples as well as those between the probationers and controls.

Downes has pointed out that a weakness of Miller's account of delinquency lies in its treatment of the non-middle-class segments of the population as a homogeneous working-class stratum.[4] An attempt to correct such an over-simple view is provided by S. M. Miller and F. Riessman.[5] They argue that a distinction needs to be made between the stable working class and the lower class, and for present purposes it is interesting to note that they suggest a greater proneness to excitement as a distinctive feature of the lower class. This view, too, is clearly congruent with those data on seeking excitement from television which were previously analysed. We have, of course, already ourselves emphasized the fact that the probationers came from a disorganized working-class stratum, and that the original matched-sample design should not be allowed to obscure the fact that the probationer and control samples probably represent two quite different social milieux.

Television and delinquency

Finally, how relevant are the findings from this study in answering questions about the effects of television on young people? What justification is there for that diffuse concern about the mass media and deviant behaviour which was mentioned at the outset?

In attempting initial replies to these questions, two provisos cannot be made too strongly. First, no survey or correlational study which merely seeks to establish whether delinquents and non-delinquents differ in their media behaviour is in a position to offer proof or disproof about the suggested causal relationships between media content and deviant or lawbreaking behaviour. Second, if such relationships exist they may well be of a more subtle sort than is commonly imagined. The whole weight of research and theory in the juvenile delinquency field would suggest that the mass media, except just possibly in the case of a very small number of pathological individuals, are never the sole cause of delinquent behaviour. At most, they may play a contributory role, and that a minor one.

In so far as the delinquents studied here did differ from the members of the control sample, then it seems proper to suggest that their particular television preferences may well have been just as much a result of their drives, needs and social position as was their delinquent behaviour itself: for example, they may have sought excitement both in the committing of delinquent acts and in living vicariously within the fantasy world of television. Of course, this does not mean to say that they necessarily sought excitement less often or less intensely in delinquency because they found it available on television. We must not make use of zero-sum models carelessly, and assume that individuals have a given and fixed magnitude of need for excitement which must be satisfied, and can be just as well satisfied in one way as another; nor, alternatively, need we necessarily assume that a taste for excitement feeds on itself.

In the present context, perhaps one of the most pertinent theories of delinquency is Sutherland's concept of differential association. Sutherland argues that "a person becomes delinquent because of an excess of definitions favourable to violation of law over definitions unfavourable to violation of law".[6] Usually, of course, the definitions favourable to lawbreaking are thought of as stemming from other persons in the immediate environment, but there is no reason why such definitions should not be provided by the mass media, including television. Indeed, this is just what those concerned about the effects of the media seek to argue in other terms. The distinctive feature of the mass media, though, is that such favourable and unfavourable definitions as they provide are equally available to all, particularly perhaps in the case of television, whereas those born into particular social groups have an access to 'personal' definitions favourable to delinquency which is totally denied to those born into other groups. Thus, so far as television can be argued to provide an over-abundance of definitions favourable to delinquency, the puzzle is why everybody does not become delinquent. Indeed, this kind of *reductio ad absurdum* is perhaps the strongest argument in favour of the suggestion that the mass media can logically at worst be only a contributory factor in the genesis of delinquency. However, in line with traditional thinking about the effects of mass communications, we must remember that evidence that the same content images are *available* to all does not mean that everybody makes use of them in the same way. It seems possible that delinquents do not merely 'use' television as a means of excitement alternative or supplementary to delinquent behaviour itself, but also seek to rationalize their delinquent acts by reference to television: if people behave aggressively and destructively on television (for example), then they may argue that it is all right for them to behave similarly. It would thus clearly be interesting in the future to collect data on delinquents' *attitudes* to the various kinds of deviant behaviour displayed on television.

In opposition to several other American theorists, Sykes and Matza have argued that it is wrong to think in terms of a specifically *delinquent* subculture.[7] Such a notion implies a social group with values fully and diametrically opposed to those of the rest of society, while empirical evidence suggests that lawbreakers accept both the legitimacy and the moral rightness of the larger social order: those who are caught manifest shame or guilt, rather than defiance. Thus the problem becomes that of accounting for transgressions of an essentially *accepted* moral and legal code. Sykes and Matza argue that the delinquent needs 'techniques of neutralization' whereby he can justify criminal acts. It seems to us that covert or conscious references to television content may well furnish one such technique. Of course, such a process of referring to mass media content is to be seen as occurring before or during the committing of delinquent acts, rather than afterwards. Apprehended delinquents who make reference to the mass media by way of rationalization are not necessarily those who have used their knowledge of the media to legitimize their behaviour in their *own* eyes earlier on.

As the interview schedule set out in Appendix A indicates, television behaviour covers a wide spectrum of relationships between viewer and medium. Only some of the areas touched on would normally be thought of as likely to elicit data relevant to the direct study of the media-delinquency relationship: differences in hours of viewing, for example, would hardly be looked on as relevant evidence unless delinquents' markedly heavier exposure could be linked to a deliberate seeking out of crime programmes. Perhaps the fact that certain aspects of the behaviour studied here can be related rather convincingly to differences in the imputed background circumstances of the three samples of adolescents underlines the fact that specific characteristics of delinquents' viewing behaviour can be accounted for in ways that have little direct bearing on the validation or disproof of simple, causal models. It is probably only through examining in detail the part which the mass media play in the lives of deviant or delinquent groups that we can arrive at better hypotheses about the patterned, and possibly, in certain senses, causal relationships that link such groups with these agencies of cultural dissemination.

Further research is needed to map more precisely the relationships between television behaviour and the particular variety of disorganized, working-class subculture which has been put forward here as characteristic of at least some of the probationers. But it seems safe to say that the research reported above could lead forward to a number of relatively precise, and testable, hypotheses. Whatever the research design employed in any future research, it would clearly now be possible to pose questions more pointed than those used in the present study. More generally it would no doubt prove fruitful to try to place delinquents' television behaviour within the context of their whole range of media behaviour and also within the context of their leisure pursuits in general. After all, delinquency itself can no doubt profitably be thought of as a form of leisure pursuit.[8]

APPENDIX A

INTERVIEW SCHEDULE (TELEVISION BEHAVIOUR)

TELEVISION RESEARCH COMMITTEE
QUESTIONNAIRE B

Area number:
Office number:
Interviewee number:
Meeting number:

Date of interview: Day of month:
Month:

Interview began at ... a.m./p.m.
Interview ended at ... a.m./p.m. } (Delete as appropriate)

1. (a) What part of an ordinary weekday—that's Monday to Friday—do you enjoy most?

   ...

   (b) Why do you enjoy that part of the day most?

   ...

   (c) If you were not you and had the chance to be someone else, who would you choose to be?

   ...

   (Use introductory statement here)

2. (a) Do you have a television where you live?

   Yes:
   No:

   (b) (Instruction: Ask only if YES to (a).)
   Do you usually watch television in the same room as the one where you normally eat your meals?

   Yes:
   No:

(c) (Instruction: Ask only if NO to (a).)
Do you ever watch television?

Yes:
No:

(d) (Instruction: Ask only if YES to (c).)
Where do you usually watch it?

3. (a) Now let's talk about reasons for watching television.
Could you tell me why you watch television?
Give me as many reasons as you can.
(Probe: Any other reasons?)

(1) ...
...
(2) ...
...
(3) ...
...
(4) ...
...

(b) Here's a list of possible reasons for watching television. Let's go through this list together; I want you to tell me if you ever watch television for any of these reasons. Of course, you can give more than one reason.
(Instruction: Let the respondent see the list while you read out each reason in turn.)

Do you ever watch television:

(1) because there's nothing else to do at the time?
(2) to get away from people?
(3) to get away from work or trouble?
(4) to have some company?
(5) because you might be able to learn something?
(6) because you might be missing something?
(7) because somebody else in the house is watching?
(8) because you know your friends will be watching?
(9) because you want to be able to talk to your friends about what's on?
(10) because you just like watching?
(11) because it's a special programme that you enjoy very much?

4. Now I'm going to read some sentences which have a word missing at the end.
I'd like you to help me to finish the sentences by giving me a word that describes your feelings. Here's an example: if I said "Having to wait for a bus always makes me feel ————," you might finish the sentence by saying "Having to wait for a bus always makes me feel impatient".
(a) "When I watch television it makes me feel
...."

(b) "I would like the television programme I watch to make me feel
...."

5. Now I'm going to read two other unfinished sentences. This time, I'd like you to give a *single* word that describes something.
Here's an example: if I said "A good word to describe most wild animals is ————", you might finish the sentence by saying "A good word to describe most wild animals is fierce".
(a) "A good word to describe most of the television programmes I actually see is
...."
(b) "A good word to describe the sort of programmes I wish were on television more often is
...."

6. (a) What sorts of television programmes do you really like? Don't give the names of actual programmes: try and give the names of *sorts* of programmes. "Sorts" means the same as "kinds" or "types". To give you the idea, if I asked you to tell me sorts of ships, you might say "liners", "sailing ships", "steamers", and so on.
(Instruction: Give no further help.)
(Probe: Any other sorts?)
(1) ...
(2) ...
(3) ...
(4) ...
(5) ...
(6) ...

(b) Now, in just the same way, what *sorts* of programme do you really *dislike*?
(Probe: Any other sorts?)
(1) ...
(2) ...
(3) ...
(4) ...

7. (a) What are some of your favourite programmes, the actual programmes you see regularly whenever you get the chance?
(Probe: Any more?)
(1) ...
(2) ...
(3) ...
(4) ...
(5) ...
(6) ...

(b) Why do you like...?
(Instruction: Mention here each of the programmes given above in turn: ask the question for all the favourite programmes given.)
(1) ...
...
(2) ...
...
(3) ...
...
(4) ...
...
(5) ...
...
(6) ...
...

8. (a) What are some of the programmes you like least, programmes that you wish weren't on television at all?
(Probe: Any more?)
(1) ...
(2) ...
(3) ...
(4) ...
(5) ...
(6) ...

(b) Why do you dislike ...?
(Instruction: Mention here each of the programmes given above in turn: ask the question for all the disliked programmes given.)

(1) ...
...
(2) ...
...
(3) ...
...
(4) ...
...
(5) ...
...
(6) ...
...

9. (a) Do you think you will probably watch some television tonight?

Yes, probably will:
Maybe, maybe not:
No:

(b) (Instruction: Ask only if NO to (a).)
Is there any special reason for that?
...

(c) Is there any special programme you plan to watch tonight?

Yes:
No:

(d) (Instruction: Ask only if YES to (c).)
Which one?
(Probe: Any others?)

(1) ...
(2) ...
(3) ...
(4) ...

(e) Do you know any other programmes that are on tonight?

Yes:
No:

(f) (Instruction: Ask only if YES to (e).)
Which ones?
(Probe: Any others?)
(1) ...
(2) ...
(3) ...
(4) ...

(g) Did you watch any television last night?
Yes:
No:

(h) (Instruction: Ask only if YES to (g).)
Which programmes did you watch?
(Probe: Any others?)
(1) ...
(2) ...
(3) ...
(4) ...

10. (a) Now I'm going to read another sentence which has a word missing at the end.
I'd like you to help me to finish the sentence.
"I think the adverts on television are ..."

(b) Here's a short list of words.
I'm going to read through the list with you.
Will you tell me all the words that describe what you really think about the adverts on television?
You can choose as many as you like.

Useful	Too long
Interesting	Helpful
Interfering	Untruthful
A nuisance	Exciting
Amusing	Too many
Welcome	A relief

(c) Would you tell me any special advert that you like?
(Instruction: Obtain as precise an identification of the particular advertisement as possible.)
...

(d) Why do you like it?
...

(e) Would you tell me any special advert that you *dis*like?
(Instruction: Obtain as precise an identification of the particular advertisement as possible.)

...

(f) Why don't you like it?

...

11. I want to find out about those hours during the week when you are likely to watch *some* television. Television programmes you see at school don't count here.
(Instruction: For each box ask a question of the form "On ordinary weekdays—that's Monday to Friday—would you be likely to see *some* television between 10 o'clock and 11 o'clock in the morning?")

Ordinary weekdays—Monday to Friday	*Saturday*	*Sunday*
10-11 a.m.	10-11 a.m.	10-11 a.m.
11-12	11-12	11-12
12- 1 p.m.	12- 1 p.m.	12- 1 p.m.
1- 2	1- 2	1- 2
2- 3	2- 3	2- 3
3- 4	3- 4	3- 4
4- 5	4- 5	4- 5
5- 6	5- 6	5- 6
6- 7	6- 7	6- 7
7- 8	7- 8	7- 8
8- 9	8- 9	8- 9
9-10	9-10	9-10
10-11	10-11	10-11
11-12	11-12	11-12

12. (a) Would you say you saw something on television

Most evenings:
3 or 4 evenings a week:
1 or 2 evenings a week:
Hardly ever:

(b) Do you usually watch television in the evening for

Less than 1 hour:
About 1 hour:
About 2 hours:
About 3 hours:
About 4 hours:
About 5 hours:
More than 5 hours:

13. (a) If you were not you and had the chance to be someone who appears on television, who would you choose to be?

...

(b) Are there any people who appear on television that you really like?
(Probe: Anybody else?)
(1) ...
(2) ...
(3) ...
(4) ...

(c) Are there any people who appear on television that you really *dis*like?
(Probe: Anybody else?)
(1) ...
(2) ...
(3) ...
(4) ...

14. Is there any particular scene, incident, special thing or special moment that you've seen on television that sticks out in your memory?
(Probe: Anything else?)
(1) ...
...
(2) ...
...

15. People sometimes talk to other people about what they've seen on television.

(a) Would you say you talked to your mother about television programmes

Very often:
Fairly often:
Not so often:
Never:

(b) Would you say you talked to your father about television programmes

Very often:
Fairly often:
Not so often:
Never:

(c) Would you say you talked to your brother(s) or sister(s) about television programmes

Very often:
Fairly often:
Not so often:
Never:

(d) Would you say you talked to somebody else in the family about television programmes

Very often:
Fairly often:
Not so often:
Never:

(e) Would you say you talked to your friend(s) or pal(s) or mate(s) about television programmes

Very often:
Fairly often:
Not so often:
Never:

(f) Would you say you talked to your girlfriend *or* boyfriend (as appropriate) about television programmes

Very often:
Fairly often:
Not so often:
Never:

(g) Would you say you talked to your teacher(s) about television programmes

Very often:
Fairly often:
Not so often:
Never:

(h) Would you say you talked to anybody we haven't mentioned so far about television programmes

Very often:
Fairly often:
Not so often:
Never:

(i) (Instruction: Ask only if *not* NEVER to (h).) Who are you thinking of?

...

16. a) Are you allowed to watch whatever you want?

Yes:
No:

(b) Are you allowed to switch the set on and off when you're on your own?

Yes:
No:

(c) Who usually decides what programme's on in your house?

...

17. Now I'd like to ask a few questions about who's with you when you watch television.

(a) When you watch television, would you say your mother's in the room

Very often:
Fairly often:
Not so often:
Never:

(b) When you watch television, would you say your father's in the room

Very often:
Fairly often:
Not so often:
Never:

(c) When you watch television, would you say your brother or sister is in the room

Very often:
Fairly often:
Not so often:
Never:

(d) When you watch television, would you say somebody else in the family is in the room

Very often:
Fairly often:
Not so often:
Never:

(e) When you watch television, would you say a friend or pal or mate is in the room

Very often:
Fairly often:
Not so often:
Never:

(f) When you watch television, would you say your girlfriend *or* boyfriend (as appropriate) is in the room

Very often:
Fairly often:
Not so often:
Never:

(g) When you watch television, would you say anybody else we haven't mentioned so far is in the room

Very often:
Fairly often:
Not so often:
Never:

(h) (Instruction: Ask only if *not* NEVER to (g).)
Who are you thinking of?
...

18. Who would you most like to watch television with?
(Instruction: Only one choice allowed.)

Mother:
Father:
Brother(s) or sister(s):
Somebody else in the family:
Girlfriend or boyfriend (as appropriate):
By yourself:
Don't know:

19. If you had a free choice and there was something you liked equally both at the pictures and on television, which would you choose?

Pictures:
Television:
Don't know:

20. (a) Can you remember a time before you ever had television where you live?
(Instruction: This does not refer to short periods, due to such things as breakdowns, when television was temporarily unavailable.)

Yes:
No:

(b) (Instruction: Ask only if YES to (a).)
How long ago was it?

Less than a year ago:
1 or 2 years ago:
3 or 4 years ago:
5 or 6 years ago:
7 or 8 years ago:
9 or 10 years ago:
More than 10 years ago:

21. What do you do in your spare time?
Forget about television for the moment.
(Instruction: Start at a new number for each separate activity.)
(Probe: Anything else?)
(1) ...
...
(2) ...
...
(3) ...
...
(4) ...
...
(5) ...
...
(6) ...
...

22. (a) Do you, or did you ever, watch television at school?

Yes:
No:

(b) (Instruction: Ask only if YES to (a).)
What did you think of it?
(Probe: Anything else?)
(1) ...
...
(2) ...
...

(Close interview)

Respondent's attitude during interview.
...
General comments on interview.
...

APPENDIX B

INTERVIEW SCHEDULES (BACKGROUND DATA)

RECORD OF SUPERVISION

Ref. No.

PART A: PERSONAL PARTICULARS, HISTORY AND ENVIRONMENT WHEN ORDER MADE

I Address | Surname | Age at date of order

First names

Date of birth | Married/Single*

Religion | Active/nominal*

II (1) Descriptive details of offence, or circumstances, leading to order (*state any offences taken into consideration or dealt with at same time*)

(2) Enquiries for Court Pretrial/Remand* Bail/Custody*

(3) Previous proceedings and court decisions with dates and courts

III (1) Probation Supervision | *Order from | to | Period of order

(2) Court making order and members

(3) Order for damages, compensation or costs

(4) Sureties

(5) Officer instituting proceedings

(6) Date | Particulars of special requirements and amendments

(7) Supervising officer | Court | Dates From | To

* *Delete word which does not apply*

TO BE COMPLETED WHEN THE CASE IS CLOSED

Date order terminated	Reason for termination

IV LIVING ACCOMMODATION

(1) Type of dwelling
(2) Structural condition
(3) Standard of furnishing
(4) Standard of cleanliness and comfort
(5) Neighbourhood—urban/rural*
(6) No. of rooms
(7) No. of persons
(8) Is there overcrowding?—yes/no*
(9) Rent per week £
Arrears (if any) at date of order £
(10) Other information

V FAMILY AND OTHER MEMBERS OF HOUSEHOLD

(a) Name and relationship (if any) to person under supervision
(b) Age, religion, occupation, average weekly earnings
(c) Health
(d) Convictions or court orders

VI PERSON UNDER SUPERVISION

(1) Whether legitimate
(2) With whom residing
(3) Use of leisure
(4) School record
(a) Name and type of school last attended
(b) Headteacher
(c) Mental ability (I.Q. if known)
(d) Educational attainment
(e) Regularity of attendance
(f) School conduct
(g) Health and physical standard
(h) Date of leaving school
(5) Employment record
(a) Employment Exchange
(b) Category
(c) Number of jobs since leaving school
(d) Employment since leaving school (give approximate dates or period and wages)

Type and Place, etc.	Remarks

VII PERSONAL HISTORIES AND ATTITUDES
of person under supervision, members of the family and other members of the household.

TELEVISION RESEARCH COMMITTEE
QUESTIONNAIRE A.C.

For all questions other than Number 6 please *circle* appropriate word or write on dotted line.

Area number: ...
Officer/interviewer number: ...
Interviewee number: ...
Group: A or B

And now to finish with, could you please give me some information about yourself and your home background? First of all,

1. Could you please give me your home address?
...
...
...
Urban:
Rural:

2. And your full name please?
(a) ...
(b) Male:
Female:
(c) Are you married or single? Married:
Single:

3. Could you please give me your age? Years ...
Months ...

4. (a) Would you say that you were a member of or belonged to any particular religion?
Yes:
No:
(b) If yes, which one, which denomination?
...
(c) Would you say that you attended church
Very often:
Fairly often:
Not so often:
Never:

5. (a) What type of house do you live in? Is it a terraced house, semi-detached, on an estate, cottage, flat, house above a shop, etc.?
...

(b) Is the house owned (by you or your parents), or is it rented?
...

(c) How many rooms are there in the house or flat where you live?

Living rooms: ...
Bedrooms: ...

6. I would now like to ask you some questions about your family.

(a) (i) Is your mother alive?
(ii) Is your father alive?
(iii) How many brothers have you?
(iv) How many sisters have you?

(b) (i) Could you please give me the age of each of your brothers, starting with the eldest and finishing with the youngest.
(ii) Could you please give me the age of each of your sisters, starting with the eldest and finishing with the youngest.

(c) (i) Does your mother live at home with you?
(ii) Does your father live at home with you?
(iii) Does your eldest brother live at home? (Repeat for each brother).
(iv) Does your eldest sister live at home? (Repeat for each sister).

(d) (i) Which of your brothers are working?
(ii) Which of your sisters are working?

(e) (i) Which of your brothers are at school?
(ii) Which of your sisters are at school?

If anyone else lives in your house/flat (household not building) apart from those just mentioned, please say who they are (uncle, grandmother, lodger, etc.).

(1) ...
(2) ...
(3) ...

7. (a) What is your father's regular job?

...

...

(b) Would you say that your father had been unemployed or out of work

Very often:
Fairly often:
Not so often:
Never:

(c) Does your mother go out to work?

Yes:
No:

(d) If yes, does she work

Part-time:
Full-time:

8. (a) Would you say your father was ill or sick, etc.

Very often:
Fairly often:
Not so often:
Never:

(b) Would you say your mother was ill or sick, etc.

Very often:
Fairly often:
Not so often:
Never:

(c) Would you say that anyone else in your family was ill or sick, etc.

Very often:
Fairly often:
Not so often:
Never:

(d) (*Instruction*: Ask only if NOT "Never" to (c).)
Who were you thinking of when you said that?

...

...

...

(End of interview for subjects still at school)

For Those Not Attending School

9. (a) Could you give me the name of the school you last attended?

...

(b) When did you leave school?

Month:
Year:

(c) You have already given me your present home address but if you lived somewhere else when you were at school (i.e. your last few weeks at school) could you also give me that address?

...
...
...

10. (a) What is your present job?

...
...

(b) How many jobs have you had since you left school?

...

(End of interview for those subjects at the club who have left school)

(Do *not* go beyond this page)

For All Interviewers

To be completed by Research Assistants from School Record Cards, Headmaster, etc.

This page should *not* be presented to or shown to the interviewee

N.B. Please follow instructions carefully in completing these questions

11. (a) Name the school that the respondent is now attending or last attended.

...

(b) Type of school:

...

(c) Name of Headteacher:
...

(d) I.Q.: ...

(e) According to the records/headmaster the educational attainment of this boy or girl is/was

In relation to class or stream {
Well above average:
Above average:
Average:
Below average:
Well below average:

(f) Number of absences over last year at school: ...
" " " " " 2 years " " : ...
" " " " " 3 years " " : ...

(9 According to the records/headmaster the conduct of this boy/girl is/was

Well above average:
Above average:
Average:
Below average:
Well below average:

(h) According to the records/headmaster the health of this boy/girl is/was

Well above average:
Above average:
Average:
Below average:
Well below average:

To be completed by Headteachers and Youth Leaders
Not to be shown to interviewees

12. In general would you say that the home background of this boy/girl is

Well above average:
Above average:
Average:
Below average:
Well below average:

13. As far as getting on with members of his/her own sex is concerned (sought after, plenty of friends, easy to get on with, etc.) would you say that he or she was

Well above average:
Above average:
Average:
Below average:
Well below average:

14. As far as getting on with members of his/her opposite sex is concerned (sought after, plenty of friends, easy to get on with, etc.) would you say that he or she was

Well above average:
Above average:
Average:
Below average:
Well below average:

15. If you know of any serious illness or disability suffered by the boy or girl or his/her family please comment:

...

...

...

16. Have you any comments on social problems or maladjustments, including illegitimacy, or any members of the family in trouble with the police, etc.?

...

...

...

17. Any further comments:

...

...

...

APPENDIX C

SAMPLING CONSIDERATIONS

Derivation of sample quotas

Since interviewing of control-sample and lower middle-class sample respondents was to be carried out at the same time as the interviewing of probationers, it was necessary to estimate the numbers of respondents needed to meet the quota requirements set up for the two other samples. Since it had already been decided that there should be only 25 per cent as many males in each control sample as there would be in the probationer sample, but the same number of females (50 per cent for females over 16), two pieces of further information were needed in order to prepare the necessary estimates of how many males and females would be required in each of the age cells in each of the control samples: (a) the total number of probationers who were likely to be interviewed (that is, the number of 'new' cases in the four areas during the specified 13-week period); and (b) the probable breakdown of this total by age and sex.

Predictions about the total number of respondents who would appear in the probationer sample were based on estimates made by each of the four Principal Probation Officers concerned about the likely figures for their own areas, derived from the caseloads of previous years. Predictions about how the grand total arrived at by aggregating the estimates from each of the four Probation Service areas should be broken down by age and sex to give estimates for individual age/sex cell requirements were to be based mainly on the published criminal statistics for 1963. The relevant statistics are reproduced in Table C1.

Table C1

*Numbers of probationers in 1963**

Age group	*Males*	*Females*	*Total*
Under 12	3,735 (10%)	341 (1%)	4,076 (11%)
12 - 13	6,166 (17%)	745 (2%)	6,911 (19%)
14 - 16	12,859 (36%)	1,700 (5%)	14.559 (41%)
17 - 20	8,732 (24%)	1,657 (5%)	10,389 (29%)
Total	31,492 (88%)	4,443 (12%)	35,935 (100%)

* Data given in *Criminal Statistics England and Wales 1963*, pp. 80-1.

In considering the success of the investigation in terms of sampling and quota requirements, therefore, there were four sets of figures to be considered:

(a) the estimated numbers of respondents which should have appeared in the various age/sex cells of the probationer sample according to the published statistics and the estimates of Principal Probation Officers;
(b) the actual numbers of respondents in the various age/sex cells of the probationer sample;
(c) quotas for the numbers of respondents in the various age/sex cells of the control and lower middle-class samples derived by taking 25 per cent, 50 per cent or 100 per cent of the age/sex cell numbers provided under (a) immediately above; and
(d) the actual numbers of respondents in the various age/sex cells of the control and lower middle-class samples.

If the age/sex cell numbers given under (b) were lower than those given under (a), then this would suggest that fewer than the total number of 'new' probationers were interviewed (assuming that the flow of cases in 1966 was roughly the same as for the preceding few years). If the age/sex cell numbers given under (d) were lower than those under (c), then this would suggest that fewer control sample respondents were recruited than should have been the case. A comparison between the age/sex cell numbers given under (d) and (b) provides a picture of the actual ratios achieved in practice in the study between the probationer sample and the two other samples. Some of these comparisons are touched on below.

On the basis of returns for previous years supplied by Principal Probation Officers, the figure of 400 cases was finally arrived at as the estimated eventual minimum overall size of the probationer sample. So as to make sure that the control and lower middle-class samples contained if anything more cases than the quota ratios called for, rather than less, a figure of 500 cases for the probationer sample was employed as the base upon which to calculate the numbers of respondents which would be needed in the two other samples.

Table C2 shows how the quota numbers for the two non-delinquent groups were derived from the 1963 criminal statistics. At each age-group level (column 1) the numbers of male probationers (column 2) and female probationers (column 3) were added together to give a figure for the total number of probationers at each age-group level (column 4). These totals were summed vertically, and the number at each age-group level expressed as a percentage of the total (column 5). These percentages were then rounded to the nearest

Table C2

*Derivation of sample quotas**

1	2	3	4	5	6	7	8	9	10	11	12	13	14	15
Age groups	*Male probationers in 1963*	*Female probationers in 1963*	*Total probationers in 1963*	*Percentage of all probationers in age groups*	*Rounded percentage of all probationers*	*500 probationers distributed by age groups*	*Females as percentage of all probationers*	*Expected nos. of female probationers*	*Rounded nos. of female probationers*	*Expected nos. of male probationers*	*Control quotas for males*	*Rounded control quotas for males*	*Control quotas for females*	*Rounded control quotas for females*
			2 + 3		R 5	6 × 5	3 − 4 × 100	7 × 8 ÷ 100	R 9	7−10	11 ÷ 4	R 12		R 14
<12	3735	341	4076	11.34	11	55	8.73	4.60	5	50	12.50	13	5	5
12-13	6166	745	6911	19.23	19	95	10.78	10.24	10	85	21.25	21	10	10
14-16	12859	1700	14559	40.51	41	205	11.68	23.94	24	181	45.25	45	24	24
17-20	8732	1657	10389	28.91	29	145	15.95	23.13	23	122	30.50	31	11.5	12
Total	31492	4443	35935	100.00	100	500	12.36		62	438		110		51

* R indicates figures rounded to the nearest whole number.

whole number (column 6). To arrive at the total numbers of probationers to be expected in each age group on the supposition that the probationer sample would include 500 individuals in all, these rounded percentages were multiplied by 5 (column 7). To partition these numbers between males and females, the percentage of females in each of the age groups was calculated by taking the number of females as a percentage of all probationers in each age group (column 3) as a percentage of column 4, yielding the estimates in column 8. To obtain an estimate of the number of female probationers at each age-group level within the total group of 500, the percentages in column 8 were taken of the numbers given in column 7. This computation gives the expected numbers of female probationers shown in column 9. These estimates were rounded to the nearest whole number (column 10). The numbers in column 10 (females) were subtracted from those in column 7 (all probationers) to give the expected numbers of male probationers in each age group supposing that the total were to be 500 (column 11). The column 11 numbers were then divided by four (25 per cent) to give the first estimate of males required for each control sample (column 12). These numbers were then rounded to the nearest whole number (column 13). This last column contains the quota requirements (by age groups) for males, for the control and lower middle-class samples. The numbers for females were derived simply by copying the entries in column 10, except for the oldest age group, where the number was halved (50 per cent). This gives the numbers in column 14. These were then rounded (column 15): the numbers here are the quota requirements (by age groups) for girls in the control and lower middle-class samples.

To get from the age-group/sex quotas shown in Table C2 to the quotas for males and females at each of the 11 age levels needed one further operation. It was, for example, necessary to partition the quota of 13 'Under 12' boys between the ages of 10 and 11.

Samples actually achieved

The statistical analysis of Table C3 shows that the *sex* distribution of cases in the probationer sample does differ from that of probationer cases from all areas in 1963. In addition, the age-group distribution for males only in the probationer sample differs from that yielded by the 1963 figures at the ·1 per cent level, and the distribution for males and females taken together differs at the same level. The age-group distribution of females is not significantly different from that of the 1963 figures. If we assume that the age-group/sex distribution

Table C3

Discrepancies between the expected and actual sex and age-group distribution of probationers

	Males			Females			Total		
Age group	*Expected*	*Actual*	*Discrepancy*	*Expected*	*Actual*	*Discrepancy*	*Expected*	*Actual*	*Discrepancy*
10 - 11	35	26	−9	3	0	−3	38	26	−12
12 - 13	57	81	+24	7	13	+6	64	94	+30
14 - 16	120	123	+3	16	18	+2	136	141	+5
17 - 20	81	51	−30	15	22	+7	96	73	−23
Total	293	281	−12	41	53	+12	334	334	0

All males and all females: $\chi^2 = 4.00$; df = 1; $p < 0.05$
Males by age: $\chi^2 = 24.11$; df = 3; $p < 0.001$
Females by age: $\chi^2 = 0.70$; df = 3; N.S.
All cases by age: $\chi^2 = 23.54$; df = 3; $p < 0.001$

of probationers was the same in the four administrative areas used in the first months of 1966 as it had been nationally three years previously (perhaps an unwarranted assumption), then it seems as though the probationer sample may well not be a representative age-group/sex quota sample of its relevant universe. Some of the youngest probationers, and particularly many of the oldest, were perhaps not interviewed (and it is easy to imagine reasons why Probation Officers should have chosen not to interview clients at these ages). This analysis certainly suggests that one should treat with extreme caution statistics descriptive of the total probationer sample as if they are to be used to construct hypotheses about even Midlands probationers in general. One might say that different sampling fractions seem to have been used at each of the age-group levels.

The probationer sample was 66 cases 'short' in terms of the original minimum expectation of 400 respondents. It is known that there are at least two major reasons for this. The most important one is that certain Probation Officers declined to take part in the investigation, so that cases coming on to the books during the 13-week period, and assigned to these officers, were not interviewed at all. It may also be that police redistricting resulted in a lower return from Staffordshire than previous years' figures suggested, with an overall 'shortage' of some 40 cases.

A full and detailed investigation was in fact carried out of the population of 'new' cases in one selected area during the relevant 13-week period. The results of this investigation are summarized in Table C4. In all, 36 relevant cases were not sent in from

Table C4

Representativeness of sample of cases from one selected area

Age group	Number of cases: *Sent in*	Number of cases: Not *sent in*
10 - 12	11 (13%)	5 (14%)
13 - 14	29 (35%)	3 (8%)
15 - 16	17 (20%)	14 (39%)
17 - 20	27 (32%)	14 (39%)
Total	84 (100%)	36 (100%)

$\chi^2 = 7.79$; df $= 2$; $p < 0.05$

this area for one reason or another; but more than this, the comparison of the two sets of cases in terms of four age groups reveals that the 84 cases sent in do not represent a fair quota of the total population in terms of the age distribution. As the analysis reported in Table C4 shows, the bias introduced by this non-representativeness was perhaps sufficient to make the total probationer sample non-representative in age-group/sex quota terms when assessed against the national figures for 1963. It will be noted in particular that the over-representation in the two lower age groups of the selected sample does serve to account for some of the general over-representation in the 12–13-year-old age group shown in Table C3.

Since data was finally available on 66 fewer probationer respondents than had originally been expected, and since the quotas for the control and lower middle-class samples were based on a probationer sample containing 500 cases (which was taken to represent the maximum possible), the overall ratios between the sizes of the non-delinquent samples and the size of the probationer sample were considerably at variance from those envisaged at the planning stage, though there was a difference between the sexes in this respect. As Table 1 in the text indicates, the lower middle-class sample of boys was about half as large as the probationer sample (instead of one quarter as large), while the control sample itself was about one third as large, or again bigger proportionately than the original research design called for. For girls, the rough equivalence between the sizes of the three samples shown in Table 1 is more nearly what was intended.

The fact that the lower middle-class sample was larger than the control sample in the case of both sexes is due to a re-allocation of respondents from the latter group to the former at the data-editing stage. It seemed clear that head teachers and youth club leaders had downgraded some respondents unfairly by allocating them to the control sample.

NOTES

Part One

Page

9 1. M. A. Elliott, 'Social Problems and Social Theories', in A. W. Gouldner and S. M. Miller (eds.), *Applied Sociology*, Collier-Macmillan, 1965, p. 401.

2. See O. N. Larsen (ed.), *Violence and the Mass Media*, Harper and Row, 1968, p. 5.

11 3. The *Guardian*, 30 March 1962, p. 1.

12 4. Sir R. Fraser, 'Violence and ITV', *Time and Tide*, 14 September 1961, pp. 1504-5.

5. Editorial comment, *Time and Tide*, 14 September 1961, p. 1499.

6. B. W. Roper, *A Ten Year View of Public Attitudes toward Television and other Mass Media 1959-1969*, Roper Research Associates, Television Information Office, March 1969, pp. 11-13.

14 7. The *Scotsman*, 15 December 1965, p. 8.

8. See P. Ferris, 'Mrs Whitehouse and the Great TV Plot', *The Observer Colour Supplement*, 10 November 1968, pp. 53-4.

9. M. Whitehouse, *Cleaning Up T.V.*, Blandford Press, 1966, pp. 23-4.

15 10. P. Barker and J. Harvey, 'Facing Two Ways: Between the 60's and 70's', *New Society*, 27 November 1969, pp. 847-50.

11. This Survey was carried out by Social Surveys (Gallup Polls) Ltd, during 1969.

16 12. This Survey was carried out by Social Surveys (Gallup Polls) Ltd. See *Polls*, vol. III, no. 3, 1968, p. 45.

18 13. F. Wertham, 'What are Comic Books?', *National Parent Teacher*, March 1949.

14. F. Wertham, 'The Comics—Very Funny', *Readers' Digest*, August 1948. Reprinted from Saturday Review of Literature.

15. F. Wertham, 'Are Comic Books Harmful to Children?' *Friends Intelligence*, 10 July 1948.

16. F. Wertham, 'What are Comic Books?', *op. cit.*

17. M. Jahoda, *The Impact of Literature: A Psychological Discussion of Some Assumptions in the Censorship Debate*, Research Centre for Human Relations, New York University, 1954. (Report prepared for the American Book Publishers Council.)

19 18. F. Wertham, 'How Movie and TV Violence Affects Children', *Ladies' Home Journal*, February 1960.

19. F. Wertham, 'School for Violence', in O. N. Larsen (ed.), *op. cit.*, p. 39.

20. F. Wertham, 'Is TV Hardening Us to the War in Vietnam?', in O. N. Larsen (ed.), *op. cit.*

20 21. A. M. Schlesinger, 'The Pornography of Violence', *The Times Saturday Review*, 4 October 1969, p. i.

21 22. For a fuller discussion, see J. D. Halloran, *Control or Consent—A Study of the Challenge of Mass Communication*, Sheed and Ward, 1963, chapter 3, pp. 53-90.

Page

22 23. M. B. Clinard, *Sociology of Deviant Behaviour*, Holt Rinehart and Winston, 1963, p. 177.

24. S. Wheeler, 'Crime and Violence', in F. T. C. Yu (ed.), *Behavioural Sciences and the Mass Media*, Russell Sage Foundation, 1968, pp. 131-50.

25. See J. D. Halloran, P. Elliott and G. Murdock, *Demonstrations and Communications—A Case Study*, Penguin Books (in the press).

26. S. Wheeler, *op. cit.*, pp. 148-9.

23 27. *Ibid.*, p. 150.

28. N. Walker, *Crime and Punishment in Britain*, Edinburgh University Press, 1965, p. 18.

24 29. D. J. West, *The Young Offender*, Penguin Books, 1967.

30. *Ibid.*, p. 34.

31. N. Walker, *op. cit.*, p. 19.

32. S. Cohen, 'The Nature of Vandalism', *New Society*, 12 December 1968, p. 875.

25 33. *Ibid.*

34. F. H. McClintock, *Crimes of Violence*, Macmillan, 1963.

35. *Criminal Statistics of England and Wales* 1968, H.M.S.O., 1969.

23 36. N. Walker, *op. cit.*, p. 36.

27 37. W. C. Reckless, *The Crime Problem*, Appleton Century Crofts, 1955, chapters 5 and 6.

38. H. Jones, *Crime in a Changing Society*, Penguin Books, 1965.

39. N. Walker, *op. cit.*

40. D. J. West, *op. cit.*

28 41. J. M. Martin and J. P. Fitzpatrick, *Delinquent Behaviour*, Random House, 1966.

42. P. B. Horton and G. R. Leslie, *The Sociology of Social Problems*, Appleton Century Crofts, 1965.

43. M. B. Clinard, *op. cit.*, pp. 177-81.

44. *Ibid.*, p. 181.

29 45. M. Jahoda, *op. cit.*, p. 19 *et seq.*

30 46. *Ibid.*, p. 11.

31 47. N. Walker, *op. cit.*, p. 105.

48. J. M. Martin and J. P. Fitzpatrick, *op. cit.*, pp. 103-40.

32 49. N. Walker, *op. cit.*, p. 107.

34 50. E. H. Sutherland and D. R. Cressey, *Principles of Criminology*, 6th edition, J. B. Lippincott, 1960, p. 211.

51. P. B. Horton and G. R. Leslie, *op. cit.*, p. 154.

52. D. Glaser, 'Criminality Theories and Behavioural Images', *American Journal of Sociology*, vol. 21, March 1956, pp. 433-45.

35 53. E. H. Sutherland, *White Collar Crime*, Dryden Press, 1949, p. 234.

54. N. Walker, *op. cit.*, p. 95.

36 55. D. Matza, *Delinquency and Drift*, John Wiley, 1964.

56. G. Sykes and D. Matza, 'Delinquency and Subterranean Values', *American Sociological Review*, vol. 26, 1961.

57. G. Sykes and D. Matza, 'Techniques of Neutralization—A Theory of Delinquency', *American Sociological Review*, vol. 22, 1957.

37 58. Irving Targoff, *Achievement Motivation, Morality and Conflict: A Study of Working Class Italian Apprentices*. Paper presented at Society for Research in Child Development, Minneapolis, 1965.

59. R. K. Merton, *Social Theory and Social Structure*, Free Press, 1957.

60. A. Cohen, *Delinquent Boys: The Culture of the Gang*, Free Press, 1955.

Page

37 61. R. A. Cloward and L. E. Ohlin, *Delinquency and Opportunity: A Theory of Delinquent Gangs*, Free Press, 1960.

39 62. *Ibid.*, p. 86.

63. D. J. West, *op. cit.*, p. 97.

64. N. Walker, *op. cit.*, p. 102-3.

40 65. D. J. West, *op. cit.*, p. 99.

66. D. M. Downes, *The Delinquent Solution*, Routledge and Kegan Paul, 1965.

67. *Ibid.*, p. 256.

68. N. H. Avison, 'The New Pattern of Crime', *New Society*, 8 September 1966, p. 310.

41 69. J. Toby, 'Economic Factors in Delinquency', in R. W. Winslow (ed.), *Juvenile Delinquency in a Free Society*, Dickenson Publishing Company, 1968, pp. 37-44.

42 70. B. Wilson, 'An Approach to Delinquency', *New Society*, 3 February 1966, pp. 8-12.

71. B. Wilson, *The Social Context of the Youth Problem* (13th Charles Russell Memorial Lecture).

72. H. A. Bloch and A. Niederhoffer, *The Gang: A Study of Adolescent Behaviour*, New York Philosophical Library, 1958, chapter 12.

73. J. F. Short, Junior, and F. L. Strodbeck, *Group Processes and Gang Delinquency*, University of Chicago Press, 1965.

43 74. W. C. Kravaceus and W. B. Miller, *Delinquent Behaviour: Culture and the Individual*, National Education Association, 1959, pp. 83-4. See also W. B. Miller, 'Lower Class Culture as a Generating Milieu of Gang Delinquency', *Journal of Social Issues*, vol. 14, 1958, pp. 5-19.

75. N. Walker, *op. cit.*, p. 103.

76. F. M. Thrasher, 'The Comics and Delinquency—Cause or Scapegoat?', *Journal of Educational Sociology*, vol. 23, 1949, pp. 195-205.

77. J. M. Martin and J. P. Fitzpatrick, *op. cit.*, p. 52.

44 78. G. M. Sykes, *Crime and Society*, Random House, 1956.

79. L. Wilkins, *Social Deviance*, Tavistock Publications, 1964.

45 80. J. R. Cavanagh, 'The Comics War', *Journal of Criminal Law and Criminology*, vol. 40, May-June 1949, pp. 34-5. Reported in Clinard, *op. cit.*, p. 179.

46 81. M. Jahoda, *op. cit.*, p. 44-5.

82. H. Blumer and P. M. Hauser, *Movies, Delinquency and Crime* (Payne Fund Studies), Macmillan, 1933.

47 83. P. G. Cressey and F. Thrasher, *Boys, Movies and City Streets* (Payne Fund Studies), Macmillan, 1933.

84. B. Gray, 'The Social Effects of the Film', *Sociological Review*, 42 (7), 1950, p. 12.

85. M. Monfredini, *The Influence of the Cinema on Juvenile Delinquency (Italy and France)*, Council of Europe, DPL/CORC (66), 2, 1966.

L. Furhammar, *The Influence of the Cinema on Juvenile Delinquency (Scandinavia and German Speaking Countries)*, Council of Europe, DPL/CORC (66), 4, 1966.

86. M. Monfredini, *op. cit.*

48 87. T. Decaigny, *Influence du cinéma sur l'enfant et l'adolescent*. Cahiers Jeunesse, Education Populaire, Bibliothèques Publiques, Brussels, 1957.

Page

48 88. G. Clostermann and K. Preuss, *Abhandlungen zur Jugend—Filmpsychologie*, Stadtisches Forscheingsinstitut für Psychologie der Arbeit und Bildung in Gelsenkirchen, Munster, 1952.

89. L. Furhammar, *op. cit.*

49 90. UNESCO, 'The Influence of the Cinema on Children and Adolescents', *Reports and Papers on Mass Communication*, no. 31, UNESCO, 1961, p. 6.

91. *Ibid.*, p. 5.

92. *Ibid.*, p. 6.

93. UNESCO, 'The Effects of Television on Children and Adolescents', *Reports and Papers on Mass Communication*, no. 43, UNESCO, 1964, 55 pp.

94. R. J. Thomson, *Television-Crime-Drama: Its Impact on Children and Adolescents*, F. W. Cheshire (Melbourne), 1959, p. 197.

95. G. Maletzke, *Television in the Life of Youth*, Hans Bredow Institute, Hamburg, 1959, 208 pp.

96. H. Himmelweit, A. N. Oppenheim, and P. Vince, *Television and the Child*, Oxford University Press, 1958, 522 pp.

50 97. T. Furu, *Television and Children's Life*, Japan Broadcasting Corporation, 1962.
Also, in English: T. Furu, 'Research on "Television and the Child" in Japan', *Studies in Broadcasting—Radio and Television*, No. 3, Culture Research Institute, NHK Japan, 1965, pp. 51-81.

98. W. Schramm, J. Lyle and E. B. Parker, *Television in the Lives of our Children*, Stanford University Press, 1961, 324 pp.

99. UNESCO (no. 43), *op. cit.*, p. 14.

100. *Ibid.*

101. E. H. Pfuhl, Junior, 'The Relationship of Mass Media to Reported Delinquent Behaviour', unpublished Ph.D. thesis, Washington State University, 1960. Published on demand by University Microfilms, High Wycombe, England.

51 102. N. Chapanis and N. C. Chapanis, 'Cognitive Dissonance Five Years Later', *Psychological Bulletin*, vol. 61, no. 1, 1964, pp. 1-22.

103. F. Wertham, 'The Scientific Study of Mass Media Effects', *American Journal of Psychiatry*, vol. 119, no. 4, October 1962, pp. 306-11.

52 104. L. A. Freedman, 'Daydream in a Vacuum Tube: A Psychiatrist's Comments on the Effects of Television', in Schramm *et al.*, *op. cit.*, pp. 189-94.

105. UNESCO (no. 43), *op. cit.*, no. 128, p. 42.

54 106. F. Emery and D. Martin, 'Psychological Effects of the "Western" Film—A Study in Television Viewing', *Studies in Mass Communication*, University of Melbourne, Department of Audio Visual Aids, 1957, 47 pp.

107. S. Feshbach, 'The Drive Reducing Function of Fantasy Behaviour', *Journal of Abnormal and Social Psychology*, vol. 50, no. 1, 1955, pp. 3-12.

108. Television and Juvenile Delinquency, *Interim Report of the Sub-Committee to Investigate Juvenile Delinquency*, U.S. Senate, 88th Congress, Second Session, 1964, 74 pp.

109. National Commission on the Causes and Prevention of Violence, *Commission Statement on Violence in Television Entertainment Programmes*, September 1969, 11 pp.

Page

54 110. F. Ferracuti and R. Lazzari, *La Violenza nei mezzi di communicazione di massa*, Radiotelevisione Italiana, 1968. Also, A. Glucksman, 'Rapport sur les Récherches, Concernment les Effets sur la Jeunesse des Scènes de Violence au Cinema et à la Television', *Communications*, July 1966, pp. 74-119.

55 111. M. W. and J. W. Riley, 'A Sociological Approach to Communication Research', in W. Schramm (ed.), *The Processes and Effects of Mass Communication*, University of Illinois Press, 1954, pp. 389-401.

56 112. L. Berkowitz, 'Violence in the Mass Media', in *Paris-Stanford Studies in Communication*, Institute for Communication Research, 1962, p. 134.

113. R. L. Hartley, *The Impact of Viewing Aggression: Studies and Problems of Extrapolation*, Office of Social Research, CBS Inc., 1964. See also R. L. Hartley, *Children vis-à-vis Film Presentation: A Review of Findings with Suggestions for Needed Research*, prepared for Joint Committee for Research on Television and Children (USA), 1966 (unpublished paper)

58 114. L. Bogart, personal communication.

115. Statement made at Assistant Masters' Conference, Southampton, 1 January 1969.

59 116. E. F. Emery and D. Martin, *op. cit.* See also L. T. Thomson, *op. cit.*

117. A. Glucksman, *op. cit.*, p. 118.

118. J. D. Halloran, 'Los efectos de la presentación, por los medios de la violencia y de la agresión', *Revista Española de la Opinión Pública*, no. 13, July-September 1968.

119. F. Ferracuti and R. Lazzari, *op. cit.*, p. 59.

60 120. J. D. Halloran, *The Effects of Mass Communication with Special Reference to Television*, Working Paper No. 1 of the Television Research Committee, Leicester University Press, 1964, p. 25.

121. The National Commission on the Causes and Prevention of Violence (USA), *Statement on Violence in Television Entertainment Programmes*, September 1969.

122. See *Violence and The Media*—A Staff Report to the National Commission on the Causes and Prevention of Violence (prepared by Robert K. Baker and Dr Sandra J. Ball), November 1969, Chapters XV, XVI, XVII.

123. *To Establish Justice to Insure Domestic Tranquility*—Final Report of the National Commission on the Causes and Prevention of Violence, December 1969, xii + 338 pp.

61 124. The National Commission on the Causes and Prevention of Violence (USA), *op. cit.*, p. 5.

125. *Ibid.*, p. 7.

126. *Ibid.*

62 127. *Ibid.*

128. Television Research Committee, *Second Progress Report and Recommendations*, Leicester University Press, 1969.

Part Two

SECTION I

Page

72 1. E. H. Pfuhl, *The Relationship of Mass Media to Reported Delinquent Behavior,* Washington State University (unpublished Doctoral dissertation), 1961, pp. 13-27.

73 2. A critical review of subcultural theories is given in D. M. Downes, *The Delinquent Solution,* Routledge and Kegan Paul, 1966, pp. 1-99.

74 3. E. H. Pfuhl, *op. cit.* See also E. H. Pfuhl, 'The Relationship of Crime and Horror Comics to Delinquency', *Research Studies of the State College of Washington,* vol. 24, 1956, pp. 170-7.

4. For a description of this method, see F. I. Nye and J. F. Short, 'Scaling Delinquent Behavior', *American Sociological Review,* vol. 22, 1957, pp. 326-31.

75 5. Writers other than Nye and Short have suggested that delinquency is widespread or even universal, though often undetected. See in particular D. Matza and G. M. Sykes, 'Juvenile Delinquency and Subterranean Values', *American Sociological Review,* vol. 26, 1961, pp. 712-19.

78 6. T. F. Hoult, 'Comic Books and Juvenile Delinquency', *Sociology and Social Research,* vol. 33, 1949, pp. 279-84.

79 7. D. K. Berninghausen and R. K. Faunce, 'An Exploratory Study of Juvenile Delinquency and the Reading of Sensational Books', *Journal of Experimental Education,* vol. 33, 1964, pp. 161-8.

SECTION II

94 1. See for example R. G. Andry, *Delinquency and Parental Pathology,* Methuen, 1960, pp. 119-26. Also G. Trasler, *The Explanation of Criminality,* Routledge and Kegan Paul, 1962, pp. 75-82.

96 2. Members of the non-probationer samples were recruited in Leicester and Leicestershire where there are many employment opportunities for women. This may partly account for such inter-sample differences as there are.

99 3. A. P. Jephcott and M. P. Carter, *The Social Background of Delinquency,* University of Nottingham (unpublished mimeographed report), 1954, pp. 122-56. For a critical analysis of this and other English studies, see D. M. Downes, *The Delinquent Solution,* Routledge and Kegan Paul, 1966, pp. 101-16.

4. W. B. Miller, 'Lower Class Culture as a Generating Milieu of Gang Delinquency', *Journal of Social Issues,* vol. 14, 1958, pp. 5-19.

SECTION III

104 1. The fullest description of such a theory is given by R. A. Cloward and L. E. Ohlin, *Delinquency and Opportunity,* Free Press, 1960.

105 2. This difference is another indication that the probationer and control samples differed in terms of subcultural behaviour patterns, or life-styles. See the discussion at the end of the previous section.

120 3. *What Children Watch,* Granada TV Network, n.d.

Page

120 4. K. Geiger and R. Sokol, 'Social Norms in Television Watching', *American Journal of Sociology*, vol. 16, 1959, pp. 174-81.

5. J. Coleman, *The Adolescent Society*, Free Press, 1961, pp. 236-43.

6. M. W. Riley and J. Riley, 'A Sociological Approach to Communications Research', *Public Opinion Quarterly*, vol. 15, 1951, pp. 445-60.

121 7. A classic paper outlining and recommending this approach is J. T. Klapper, 'Mass Communication Research: An Old Road Resurveyed', *Public Opinion Quarterly*, vol. 27, 1963, pp. 515-27.

SECTION IV

128 1. J. B. Mays, *Growing Up in the City*, Liverpool University Press, 1954.

129 2. *Ibid.*, p. 105. See S. Glueck and E. Glueck, *Unravelling Juvenile Delinquency*, Commonwealth Fund, 1950, p. 162.

3. Mays, *op. cit.*, p. 106.

4. W. B. Miller, 'Lower Class Culture as a Generating Milieu of Gang Delinquency', *Journal of Social Issues*, vol. 14, 1958, pp. 5-19.

131 5. L. Bailyn, 'Mass Media and Children: A Study of Exposure Habits and Cognitive Effects', *Psychological Monographs*, vol. 71, 1959, pp. 1-48.

132 6. Glueck and Glueck, *op. cit.*, pp. 244-53.

133 7. T. Parsons, 'Certain Primary Sources and Patterns of Aggression in the Social Structure of the Western World', *Psychiatry*, vol. 10, 1947, pp. 167-81.

134 8. Miller, *op. cit.*, p. 9.

SECTION V

142 1. Perhaps this difference reflects an objective situation, and probationers had actually had fewer opportunities for ETV viewing. Alternatively, this may be a matter of selective memory: among the less enthusiastic probationers, more *forgot* that they had seen ETV.

143 2. A. K. Cohen, *Delinquent Boys: The Culture of the Gang*, Routledge and Kegan Paul, 1956, pp. 132-7.

144 3. This is similar to the notion of an over-rejection of identification with the mother.

SECTION VI

147 1. R. G. Andry, *Delinquency and Parental Pathology*, Methuen, 1960.

2. *Ibid.*, pp. 40-59.

150 3. Available data from other studies tend to show that delinquents are no more isolated than matched controls.

152 4. In the light of previous discussions of the delinquents' view of their sex role, it must be emphasized that they were perhaps not *trying* to 'get on' with girlfriends.

156 5. J. Klein, *Samples from English Cultures*, Routledge and Kegan Paul, 1965, vol. 1, p. 27 *et passim*. For the original 'Ashton' study, see N. Dennis, F. Henriques and C. Slaughter, *Coal Is Our Life*, Eyre and Spottiswoode, 1956.

157 6. B. Bernstein, 'Social Class and Linguistic Development', in A. H. Halsey, J. Floud and N. Anderson (eds.), *Education, Economy and Society*, Free Press, 1961, pp. 288-314

7. Klein, *op. cit.*, pp. 517-26.

160 8. *Ibid.*, p. 536.

Page

160 9. For a discussion, see J. T. Klapper, *Effects of Mass Communication*, Free Press, 1960, pp. 26-37.

SECTION VII

168 1. B. Sugarman, 'Involvement in Youth Culture, Academic Achievement and Conformity in School', *British Journal of Sociology*, vol. 18, 1967, pp. 151-64.

169 2. Downes offers a rather different account of the relationships between delinquency and youth culture. See D. M. Downes, *The Delinquent Solution*, Routledge and Kegan Paul, 1966, pp. 129-34.

3. Sugarman, *op. cit.*, p. 160.

SECTION VIII

173 1. D. M. Downes, *The Delinquent Solution*, Routledge and Kegan Paul, 1966, p. 100.

176 2. W. L. Schramm, J. Lyle and E. B. Parker, *Television in the Lives of our Children*, Stanford University Press, 1961, pp. 98-117.

177 3. W. B. Miller, 'Lower Class Culture as a Generating Milieu of Gang Delinquency', *Journal of Social Issues*, vol. 14, 1958, pp. 5-19.

178 4. Downes, *op. cit.*, pp. 71-3.

5. S. M. Miller and F. Riessmann, 'The Working Class Subculture: A New View', *Social Problems*, vol. 9, 1961, pp. 86-97.

179 6. E. H. Sutherland, *Principles of Criminology*, J. B. Lippincott, 1939, p. 6.

180 7. G. M. Sykes and D. Matza, 'Techniques of Neutralization: A Theory of Delinquency', *American Sociological Review*, vol. 22, 1957, pp. 664-70.

181 8. D. Matza and G. M. Sykes, 'Juvenile Delinquency and Subterranean Values', *American Sociological Review*, vol. 26, 1961, pp. 712-19.

INDEX OF NAMES

This index refers only to names specifically mentioned in the text

SUBJECT INDEX